Lecture Notes in Mathematics 2085

For further volumes:
http://www.springer.com/series/304

Arnaud Debussche • Michael Högele
Peter Imkeller

The Dynamics of Nonlinear Reaction-Diffusion Equations with Small Lévy Noise

Springer

Arnaud Debussche
Antenne de Bretagne
Ecole Normale Supérieure Cachan
Bruz, France

Michael Högele
Institut für Mathematik
Universität Potsdam
Potsdam, Germany

Peter Imkeller
Institut für Mathematik
Humboldt-Universität zu Berlin
Berlin, Germany

ISBN 978-3-319-00827-1 ISBN 978-3-319-00828-8 (eBook)
DOI 10.1007/978-3-319-00828-8
Springer Cham Heidelberg New York Dordrecht London

Lecture Notes in Mathematics ISSN print edition: 0075-8434
ISSN electronic edition: 1617-9692

Library of Congress Control Number: 2013944220

Mathematics Subject Classification (2010): 60H15; 60G52; 60G18; 35K57

Printed on acid-free paper

Springer is part of Springer Science+Business Media (www.springer.com)

Preface

Dynamical systems perturbed by small random noise have received a vast attention over the last decades in many areas of science extending from physics through chemistry and biology to climatology. They typically represent a deterministic large scale phenomenon expressed in terms of an ordinary or partial differential equation which inherits the noisy residual of a rapidly fluctuating low intensity perturbation on much smaller scales. Commonly, these systems largely mimic the phenomenon's unperturbed deterministic behavior up to a characteristic time scale. This scale is a function of the intensity of the perturbation, depends essentially on the underlying nature of the noise and, to a minor extent, on the state space geometry of the deterministic system. Beyond that scale the system exhibits noise induced excursions.

If the deterministic system has several stable equilibria to which it converges in generic relaxation times if started in their respective domains of attraction, these excursions lead to transitions between different equilibria starting from small neighborhoods of one of them. If the system is rescaled with its characteristic time scale, the quasi-deterministic waiting time for a transition from an initial equilibrium is of the order of a time unit on an exponential clock. In its characteristic time scale, the complex fluctuating perturbed system therefore behaves asymptotically as a continuous time Markov chain switching between the stable equilibria of the unperturbed system, turning them into *metastable* states.

In the mathematics literature such systems first appeared in the beginning of the 1970s, mainly in the context of large deviations for Gaussian perturbations. For this type of noise, characteristic time scales are of order $\exp(V/\varepsilon^2)$, where ε is the noise intensity, and the quantity V related to the geometry of the deterministic system. The large deviations approach as well as its potential theoretic extension turned out to be very fertile, and large deviation principles describing their metastable behavior have been discovered for large classes of ordinary and partial differential equations.

For dynamical systems with non-Gaussian noise, exit and transition problems have been much less studied. The most interesting non-Gaussian noise is given by the α-stable one, arising in local limit theorems for heavy-tailed random walks. The most prominent example in this class is Cauchy noise, well known to lack first moments as well as a suitable Cameron–Martin space. Therefore for the study of the metastable behavior of dynamical systems perturbed by it, large deviation techniques may not apply. After an abstract approach via its Markov generator by Godovanchuk in 1979, Imkeller and Pavlyukevich solved the first exit problem for one-dimensional systems and described their metastable behavior in 2006. Their study is crucially based on a skilled distinction between large and small jumps of the noise and the strong Markov property of the system, which allows to compensate for the lack of moments. The precise heuristics behind this approach is explained in detail in Sect. 1.2. In strong contrast to the Gaussian case, the characteristic time scale is of order Q/ε^{α}, where ε is the noise intensity, α the stability index of the noise, and Q a quantity depending on the geometry of the deterministic system and the Lévy measure.

These lecture notes treat the first exit problem and metastability for a paradigm class of reaction–diffusion equations—the Chafee–Infante equations—perturbed by additive regularly varying noise in the infinite-dimensional space of weakly differentiable functions over an interval. The corresponding principal results are contained in the following theorems. Theorem 5.16 states the convergence of the rescaled first exit times from domains of attraction of equilibria to those of a reduced model in terms of exponential moments on the same probability space. Theorem 7.10 describes metastability for the system in the characteristic time scale. To our knowledge this is the first treatment of this type of problems for stochastic partial differential equations. Also the techniques used in the proofs are new to the field.

The lecture notes address graduate students and researchers in mathematics and natural scientists with a background in partial differential equations and stochastic analysis, who would like to understand in detail the rich and subtle interplay of the deterministic infinite-dimensional dynamics and the jump behavior in terms of the Lévy measure of the random perturbation.

The text is as self-contained as possible with a proof or at least a sketch of it for every proposition in all different areas involved. In particular we give an overview of the literature on the deterministic Chafee–Infante equations. We prove fine estimates on the relaxation time in Chap. 2, which do not exist in the literature so far. In the sequel we give an introduction to stochastic reaction–diffusion equations and establish all properties relevant to our purposes, in particular the existence of a global solution and the strong Markov property in Chap. 3. The mathematical core of the text is presented in Chaps. 4–7. It concludes with an additional chapter in the appendix, where we explain the climate dynamics motivation for our paradigm model.

Bruz, France — Arnaud Debussche
Potsdam, Germany — Michael Högele

Berlin, Germany — Peter Imkeller
April 2013

Acknowledgments

The authors express their gratitude to Berlin Mathematical School (BMS) for financial and infrastructure support, Ecole Normale Supérieure Cachan, Antenne de Bretagne, for the kind hospitality during a stay of the second author in the winter of 2008/2009, and Ilya Pavlyukevich for many valuable discussions.

Contents

Notation

Important Constants

- $\alpha \in (0, 2)$, index of the noise, see ν and L
- $\rho \in (\frac{1}{2}, 1)$, see $\frac{1}{\varepsilon^\rho}, \varepsilon \in (0, 1)$
- $\Gamma > 0$, large geometric constant
- $\gamma > 0$, appropriately small exponent

The Spaces

- $(L^2(0, 1), |\cdot|)$, Lebesgue space of equivalence classes of square integrable functions on $(0, 1)$ with the usual norm
- $|\cdot|_p$, $p \neq 2$, the norm of the Lebesgue space $L^p(0, 1)$
- $H = H_0^1(0, 1)$, $(H, \|\cdot\|)$, space of weakly differentiable elements of $L^2(0, 1)$ with Dirichlet boundary conditions with $\nabla x \in L^2(0, 1)$ for $x \in H$ and with the norm $\|x\|^2 = \int_0^1 (\nabla x(\zeta))^2 d\zeta$, $x \in H$
- $B_r(x)$, $x \in H$, $r > 0$ is the ball in H of center x and radius r.
- $(\mathscr{C}_0(0, 1), |\cdot|_\infty)$, space of continuous functions on $[0, 1]$ with $x(0) = x(1) = 0$ with the supremum norm
- $D(\mathbb{R}^+; H)$, space of càdlàg functions on $\mathbb{R}^+ = [0, \infty)$ with values in H

The Deterministic Chafee–Infante Equation

- $(S(t))_{t \geqslant 0}$, heat semigroup on H
- $\lambda > \pi^2$, with $\lambda \neq (k\pi)^2$, $k \in \mathbb{N}$, Chafee–Infante parameter
- $u = (u(t; x))_{\substack{t \geqslant 0 \\ x \in H}}$, solution of the deterministic Chafee–Infante equation at time $t \geqslant 0$ with initial value $x \in H$ for fixed parameter λ
- $v_\theta = (v_\theta(t; x))_{\substack{t \geqslant 0 \\ x \in H}}$, solution of the deterministic Chafee–Infante equation with nonlinearity $f(\cdot + \theta(t))$ at time $t \geqslant 0$ with initial value $x \in H$ for fixed parameter λ and $\theta \in L^\infty(0, \infty; H)$
- $\phi^\pm$, one of the two stable states $\{\phi^+, \phi^-\}$ of u for fixed λ
- $\mathscr{A}^\lambda$, global attractor of the dynamical system $t \mapsto u(t; \cdot)$ in H for fixed λ

Domains of Attraction

Let $\delta_i > 0, i = 1, 2, 3$, and $\varepsilon, \gamma \in (0, 1)$.

- $D^{\pm}$, domain of attraction of $\phi^{\pm}$ under the flow $t \mapsto u(t; x), x \in H$
- $\mathscr{S} := H \setminus (D^+ \cup D^-)$, smooth manifold separating D^+ and D^-, called separatrix
- $D^{\pm}(\delta_1) := \{x \in D^{\pm} \mid \cup_{t \geqslant 0} B_{\delta_1}(u(t; x)) \in D^{\pm}\}$
- $D^{\pm}(\delta_1, \delta_2) := \{x \in D^{\pm} \mid \forall\, \theta \in \mathbb{D}(\mathbb{R}^+; H)$ with $\sup_{t \geqslant 0} \|\theta(t)\| \leqslant \delta_2 :$
  $$\bigcup_{t \geqslant 0} B_{\delta_2}(v_\theta(t; x)) \in D^{\pm}(\delta_1)\}$$
- $\tilde{D}^{\pm}(\varepsilon^\gamma) := D^{\pm}(\varepsilon^\gamma, \varepsilon^{2\gamma})$
- $\tilde{D}^{0}(\varepsilon^\gamma) := H \setminus (\tilde{D}^{+}(\varepsilon^\gamma) \cup \tilde{D}^{-}(\varepsilon^\gamma))$
- $\tilde{D}^{\pm 0}(\varepsilon^\gamma) := \tilde{D}^{+}(\varepsilon^\gamma) \bigcup \tilde{D}^{0}(\varepsilon^\gamma)$
- $D^{\pm}(\delta_1, \delta_2, \delta_3) := \{x \in D^{\pm} \mid \forall\, \theta \in D(\mathbb{R}^+; H)$ with $\sup_{t \geqslant 0} \|\theta(t)\| \leqslant \delta_3 :$
  $$\cup_{t \geqslant 0} B_{\delta_3}(v_\theta(t; x)) \in D^{\pm}(\delta_1, \delta_2)\}$$
- $\hat{D}^{\pm}(\varepsilon^\gamma) := D^{\pm}(\varepsilon^\gamma, \varepsilon^{2\gamma}, \varepsilon^{2\gamma})$
- $D^{\pm}(\delta_1, \delta_2, \delta_3, \delta_4) := \{x \in D^{\pm} \mid B_{\delta_4}(x) \in D^{\pm}(\delta_1, \delta_2, \delta_3)\}$
- r^*, radius of a ball such that all $v_\theta(\cdot; x)$ enters this ball in a time independent of $x \in H$ and θ, $\sup_{t \geq 0} \|\theta(t)\| \leq 1$, $B_{r^*}(0)$ absorbing set of u
- s_{r^*}, uniform bound from below on the time
- $T_{rec} + \kappa\gamma|\ln \varepsilon|$, upper bound for $u(t; x)$, $x \in D^{\pm}(\varepsilon^\gamma)$, to enter $B_{(1/2)\varepsilon^{2\gamma}}(\phi^{\pm})$

Shifted Domains of Attraction

Let $\delta_i > 0, i = 1, 2, 3$, and $\varepsilon, \gamma \in (0, 1)$.

- $D_0^{\pm} = D^{\pm} - \phi^{\pm}$
- $D_0^{\pm}(\delta_1) = D^{\pm}(\delta_1) - \phi^{\pm}$
- $D_0^{\pm}(\delta_1, \delta_2, \delta_3, \delta_4) = D^{\pm}(\delta_1, \delta_2, \delta_3, \delta_4) - \phi^{\pm}$
- $\tilde{D}_0^{\pm}(\varepsilon^\gamma) = \tilde{D}^{\pm}(\varepsilon^\gamma) - \phi^{\pm}$
- $\hat{D}_0^{\pm}(\varepsilon^\gamma) = \hat{D}^{\pm}(\varepsilon^\gamma) - \phi^{\pm}$
- $\hat{D}^{0}(\varepsilon^\gamma) = H \setminus (\hat{D}^{+}(\varepsilon^\gamma) \cup \hat{D}^{-}(\varepsilon^\gamma))$

The Stochastic Chafee–Infante Equation

- $\varepsilon \in (0, 1)$, noise intensity
- ν, symmetric, regularly varying Lévy measure on $\mathscr{B}(H)$ of index $\alpha \in (0, 2)$
- $L = (L(t))_{t \geqslant 0}$, symmetric pure jump Lévy process in H with Lévy measure ν
- $X^\varepsilon = (X^\varepsilon(t; x))_{t \geqslant 0}$, solution of the stochastic Chafee–Infante equation driven by εdL at time $t \geqslant 0$ with initial value $x \in H$
- $\Delta_t L = L(t) - L(t-)$, jump of L at time $t > 0$
- $\dfrac{1}{\varepsilon^\rho}$, for $\varepsilon, \rho \in (0, 1)$, jump height threshold of L between "small" and "large" jumps
- $\eta^\varepsilon = (\eta^\varepsilon(t))_{t \geqslant 0}$, compound Poisson process consisting of all jumps of L of height $\|\Delta_t L\| > \frac{1}{\varepsilon^\rho}$, called "large" jumps

- $(T_i)_{i\in\mathbb{N}}$, jump times of η^ε
- $t_i = T_i - T_{i-1}, i \in \mathbb{N}$, waiting times between jumps the of η^ε
- $W_i = \Delta_{T_i} L, i \in \mathbb{N}$, i-th jump (increment) of η^ε
- $\xi^\varepsilon = (\xi^\varepsilon(t))_{t\geqslant 0}$, where $\xi^\varepsilon(t) = L(t) - \eta^\varepsilon(t), t \geqslant 0$, called "small" jumps process
- $\xi^* = (\xi^*(t))_{t\geqslant 0}$, where $\xi^*(t) = \int_0^t S(t-s)\mathrm{d}\xi^\varepsilon(s)$, called "small" jumps convolution
- $Y^\varepsilon = (Y^\varepsilon(t;x))_{\substack{t\geqslant 0,\\ x\in H}}$, mild solution of the stochastic Chafee–Infante equation driven by $\varepsilon\mathrm{d}\xi^\varepsilon$ at time $t \geqslant 0$ and initial value $x \in H$

Time Scales

Let $\varepsilon > 0$, $\rho \in (\frac{1}{2}, 1)$, $\alpha \in (0, 2)$, and write $f_\varepsilon \approx_\varepsilon g_\varepsilon$ for $\lim_{\varepsilon\to 0+} f_\varepsilon/g_\varepsilon = 1$.

- $\lambda^\pm(\varepsilon) = \nu\left(\frac{1}{\varepsilon}(D_0^\pm)^c\right) \approx_\varepsilon \varepsilon^\alpha\ \ell(1/\varepsilon)\ \mu\left((D_0^\pm)^c\right)$, characteristic rate of the first exit time
- $\beta_\varepsilon = \nu\left(\frac{1}{\varepsilon^\rho}B_1^c(0)\right) \approx_\varepsilon \varepsilon^{\alpha\rho}\ \ell(1/\varepsilon^\rho)\ \mu\left(B_1^c(0)\right)$, intensity of η^ε
- $\lambda^0(\varepsilon) = \nu\left(\frac{1}{\varepsilon}B_1^c(0)\right) \approx_\varepsilon \varepsilon^\alpha\ \ell(1/\varepsilon)\ \mu\left(B_1^c(0)\right)$, characteristic rate of metastability
- $\ell : (0,\infty) \to (0,\infty)$, slowly varying function associated with ν
- μ, limit measure of ν on $\mathscr{B}(H)$

Exit Times and Transition Times

Let $\varepsilon, \gamma \in (0, 1)$.

- $\tau_x^\pm(\varepsilon)$, first exit time of $X^\varepsilon(\cdot;x), x \in \hat{D}^\pm(\varepsilon^\gamma)$ from the reduced domain of attraction $\tilde{D}^\pm(\varepsilon^\gamma)$
- $\tau_x^{\pm 0}(\varepsilon)$, first exit time of $X^\varepsilon(\cdot;x), x \in \hat{D}^\pm(\varepsilon^\gamma)$ from the enhanced domain of attraction $\hat{D}^{\pm 0}(\varepsilon^\gamma)$
- $\chi_x^\pm(\varepsilon)$, first entrance time of $X^\varepsilon(\cdot;x), x \in \hat{D}^\pm(\varepsilon^\gamma)$ in $B_{\varepsilon^{2\gamma}}(\phi^\mp)$
- $\tau_x^0(\varepsilon)$, first exit time from the neighborhood of the separatrix $\hat{D}^0(\varepsilon^\gamma)$

Chapter 1
Introduction

1.1 Motivation of the Exit Time Problem from Climate Dynamics

Our primary interest in this book lies in the study of dynamical properties of reaction-diffusion equations perturbed by Lévy noise of intensity ε in the small noise limit $\varepsilon \to 0$. The material of the book is based on the Ph.D. thesis [Hög11] by M. Högele. Typically, a reaction diffusion equation we consider is supposed to possess two domains of attraction connected by a separating manifold. Without perturbations by noise, the system's solution trajectories would relax to the stable equilibrium of the domain of attraction in which they are started. If noise is turned on, spontaneous transitions from one domain of attraction to the other one become possible, through large deviations of the noisy system in the Gaussian case, and eventually through jumps in the case of more general Lévy noise. In any case, additive noise transforms the stable states in the domains of attraction into *metastable* ones with characteristic transition times depending on the noise amplitude. One of the main problems we shall address is concerned with describing the asymptotic order of time as a function of noise amplitude ε it takes the system to switch from one domain of attraction to the other one—or from one *metastable regime* to the other one—in the small noise limit $\varepsilon \to 0$. In the Gaussian case, the transition dynamics has been intensively studied and well understood mainly on the basis of the Freidlin–Wentzell theory of noisy perturbations of dynamical systems. As will become clear below, in the case of non-Gaussian Lévy noise, this involves detailed and subtle estimates on the time the system will spend in neighborhoods of the separating manifold. Primarily for this reason, we chose to restrict our attention on one particular class of reaction-diffusion equations, the Chafee–Infante equation described in more detail below. As one of its main features, the Chafee–Infante equation exhibits two domains of attraction connected by a smooth separating manifold the globally complex structure of which is well understood. This will enable us to assess questions about residence times in its small

A. Debussche et al., *The Dynamics of Nonlinear Reaction-Diffusion Equations with Small Lévy Noise*, Lecture Notes in Mathematics 2085, DOI 10.1007/978-3-319-00828-8_1,

neighborhoods to a degree that suffices to derive the global features of the dynamics of transitions. The need to have a more detailed understanding of the meandering of trajectories of the noisy system near parts of a complex separating manifold is the only reason for us to confine our study to this particular class of reaction-diffusion equations with two domains of attraction. We are confident that our general line of reasoning applies to a much more general class of reaction-diffusion equations for instance with finitely many domains of attraction. The main obstacle to overcome in a generalization consists in formulating conditions on the noise which guarantee that the system does not get caught for too long in neighborhoods of manifolds separating domains of attraction the structure of which should be sufficiently well described for this purpose. We refrain from formulating such conditions here, and leave generalizations to other systems of reaction-diffusion equations for further research. Our initial motivation to look for problems of this kind originates in a climate dynamics context. Roughly, the two domains of attraction have to be interpreted as two stable climate states in a conceptual energy balance type climate model. In a noisy environment, they describe metastable states of the global averaged temperature, typically cold and warm states. The guiding question asked concerns typical times for transitions between them triggered by noise.

Let us introduce the main object of our study, the Chafee–Infante equation perturbed by Lévy noise, as one of the simplest idealized semilinear stochastic reaction-diffusion equations. Of course, the asymptotic study of its dynamics in the small noise limit possesses interest independently of any particular background in which it may arise. Some of the intuition behind its main terms will be motivated by briefly looking at this simple climate dynamics background.

Noisy energy balance models aim at describing qualitative features of the global temperature, seasonally and longitudinally averaged, as a function $X^{\varepsilon}(t, \zeta)$ of time and the zonal position ζ identified with a point on the unit interval, perturbed by spatial–temporal noise of (small) intensity $\varepsilon > 0$. The underlying temporal evolution of temperature on the interval $[0, 1]$ limited by the poles involves random processes taking their real values in sets of functions on compact domains. This leads directly to equations in infinite-dimensional spaces, and infinite-dimensional models of noise, formally to an SPDE. In the light of our guiding example, its three components may be interpreted in the following way.

1. A *reaction* term f of the evolution equation may be seen as expressing a deterministic forcing of temperature. It derives heuristically from simple assumptions on the balance between absorbed and emitted solar radiation energy as a function of time (see [Imk01]). Absorbed energy is qualified as a function of the temperature dependent albedo function, and emitted energy by the Stefan–Boltzmann law for black body radiators as being proportional to the forth power of temperature. The resulting energy balance as a function of temperature has two stable and one unstable zero representing equilibria of a dynamical system. Hence the resulting reaction term can be described as the negative gradient of a potential function $f = -U'$ with two local minima representing a cold and a warm basic climate state.

2. A spatial *diffusion term* $\frac{\partial^2}{\partial \zeta^2} X^\varepsilon$ may be seen in our model motivation as representing heat diffusion between equator and poles which is caused by different rates of insolation due to different angles of incidence of sunlight.
3. The energy balance based reaction term and the heat diffusion term lead—in an idealized version—to a deterministic Chafee–Infante equation. According to Hasselmann's approach (see Arnold [Arn01] and Hasselmann [Has76]) this equation may be seen to be perturbed by an *additive stochastic process* L of small intensity $\varepsilon > 0$ taking values in an appropriate function space on the interval $[0, 1]$. It represents unresolved solar and atmospheric forcing. Following the suggestion in Ditlevsen [Dit99] and Gairing et al. [GHIP11] we may take L to be of Lévy type with jump measure tails of polynomial order. The most prominent example is the case of α-stable noise.

With this motivating example in mind, let us now turn to the investigation of the dynamics of the Chafee–Infante equation from a general perspective, in particular its exit and transition dynamics between the domains of attraction of the metastable states. We will denote the solution of the deterministic Chafee–Infante equation by $u = X^0$. It formally satisfies

$$\begin{aligned} \frac{\partial}{\partial t} u(t,\zeta) &= \frac{\partial^2}{\partial \zeta^2} u(t,\zeta) + f(u(t,\zeta)), \ \zeta \in [0,1], \ t > 0, \\ u(t,0) &= u(t,1) = 0, \qquad\qquad t > 0, \\ u(0,\zeta) &= x(\zeta), \qquad\qquad \zeta \in [0,1], \end{aligned} \tag{1.1}$$

where $U(y) = (\lambda/4)y^4 - (\lambda/2)y^2$ for $\lambda > 0$ fixed, and $f = -U'$.

The solution takes values in an infinite-dimensional function space, as for example $L^2(0,1)$, $H_0^1(0,1)$ or $\mathscr{C}_0(0,1)$, where also the initial state x is taken (see [Tem92] or [SY02]). Since its pure reaction term f has two zeros given by the minima of U, apart from singular values of λ, the Chafee–Infante equation possesses in a generic setting two hyperbolic stable states $\phi^+, \phi^- \in \mathscr{C}^\infty(0,1)$. Nevertheless, there may be several unstable saddles, depending on the value of the parameter λ.

If the additive Lévy noise term of intensity $\varepsilon > 0$ is added as a perturbation to the deterministic equation, we obtain the stochastic Chafee–Infante equation of the form

$$\begin{aligned} \frac{\partial}{\partial t} X^\varepsilon(t,\zeta) &= \frac{\partial^2}{\partial \zeta^2} X^\varepsilon(t,\zeta) + f(X^\varepsilon(t,\zeta)) + \varepsilon \dot{L}(t,\zeta), \ \zeta \in [0,1], \ t > 0, \\ X^\varepsilon(t,0) &= X^\varepsilon(t,1) = 0, \qquad\qquad t > 0, \\ X^\varepsilon(0,\zeta) &= x(\zeta), \qquad\qquad \zeta \in [0,1], \end{aligned} \tag{1.2}$$

where $\lambda > 0$ and $f = -U'$. The noise term $\dot{L}$ formally represents the generalized derivative of a pure jump Lévy process in the Sobolev space $H = H_0^1(0,1)$ with Dirichlet boundary conditions, regularly varying Lévy measure of index $\alpha \in (0,2)$ and initial value $x \in H$.

For the one-dimensional counterpart of (1.2) without diffusion term Imkeller and Pavlyukevich investigate the asymptotic behavior of exit and transition times in the small noise limit in [IP06a, IP08] and [IP06b]. In contrast to the Wiener case, for which exponential growth with respect to the noise intensity is observed in [FV98], these models feature exit rates with polynomial growth in the limit of small noise. Accordingly, the critical time scale in which the global metastable behavior of the jump diffusion can be reduced to a finite state Markov chain jumping between the metastable states (see also [BEGK04]) is equally polynomial in the noise intensity.

In this book we shall be primarily concerned with the question: *To which extent do these results still hold true in the infinite dimensional Chafee–Infante reaction-diffusion framework, with corresponding infinite-dimensional noise?*

We shall show in Theorem 5.11 that the expected exit time from (reduced) domains of attraction of the metastable states ϕ^+, ϕ^- increases polynomially of order $\varepsilon^{-\alpha}$ in the limit of small noise intensity ε, and characterize the exit scenarios. We shall also show in Theorem 7.10 that for this time scale of ε the jump diffusion system reduces to a finite state Markov chain with values in the set of stable states $\{\phi^+, \phi^-\}$. Our analysis can be considered as a starting point for studying metastable behavior of dynamical systems induced by reaction-diffusion equations perturbed by Lévy jump noise on a more general basis. We also note that our model gives rise to order preserving random dynamical systems (see [Chu01]). This property potentially has in store further information on qualitative asymptotic behavior of the system, for instance on the structure of its pullback attractors.

1.2 Heuristics for the First Exit Times: Noise Decomposition into Small and Large Jumps

The study of exit times from domains of attraction will be the main ingredient of our investigation of the dynamical properties of the Chafee–Infante equation perturbed by Lévy noise. In this section we explain the heuristics of the method to determine the expected first exit time for a domain of attraction of the stable states $\phi^\pm$ in the asymptotics of small noise intensity. In doing this, we extend the arguments given in [IP08] for dimension 1 which proceed along the following lines.

Step 1. A detailed study of the stable solutions as well as the separating manifold of the deterministic Chafee–Infante equation leads to the construction of reduced versions $D^\pm(\varepsilon^\gamma) \subset D^\pm$ of the domains of attraction $D^\pm$ of the stable solutions $\phi^\pm$ such that the solution $u(t;x)$ of the Chafee–Infante equation starting in $x \in D^\pm(\varepsilon^\gamma)$ finds itself within a small neighborhood $B_{\varepsilon^{2\gamma}}(\phi^\pm)$ at times t exceeding $T_{rec} + \kappa\gamma|\ln\varepsilon|$. Here T_{rec} is a global relaxation time and $\kappa > 0$ a global constant, formally

$$u(t;x) \in B_{\varepsilon^{2\gamma}}(\phi^\pm) \qquad \text{for all} \quad t \geqslant T_{rec} + \kappa\gamma|\ln\varepsilon| \quad \text{and} \quad x \in D^\pm(\varepsilon^\gamma). \tag{1.3}$$

Step 2. For a threshold $c > 0$ we recursively define the sequence of jump times of the driving Lévy process L with values in H exceeding c by

$$T_{i+1} := \inf\{t > T_i \mid \|\Delta_t L\| > c\}, \qquad T_0 = 0,$$

where for $t \geqslant 0$ and a process Y we write $\Delta_t Y = Y(t) - Y(t-)$. If $(S(t))_{t\geqslant 0}$ is the Markovian semigroup associated with the diffusion operator on $(0, 1)$, and we use the mild solution formulation following [PZ07], the jumps of X^ε are just the jumps of L, i.e.

$$\Delta_{T_i} X^\varepsilon = \Delta_{T_i} \int_0^{\cdot} S(\cdot - s)\mathrm{d}L(s) = \Delta_{T_i} L, \qquad i \in \mathbb{N}. \tag{1.4}$$

We let the threshold c depend on ε, and choose $c = c(\varepsilon) = \frac{1}{\varepsilon^\rho}$ for $\rho \in (0, 1)$ to split $L(t) = \xi^\varepsilon(t) + \eta^\varepsilon(t)$ into a small jump part ξ^ε, with

$$\varepsilon\|\Delta_t \xi^\varepsilon\| \leqslant \varepsilon \frac{1}{\varepsilon^\rho} \to 0, \quad \varepsilon \to 0+ \tag{1.5}$$

and a large jump part η^ε, with $\eta^\varepsilon(t) = \sum_{i:T_i \leqslant t} \Delta_{T_i} L$, $t \geqslant 0$. Between two large jump times T_i and T_{i+1}, the strong Markov property allows us to consider X^ε as being driven by the small jump component $\varepsilon\xi^\varepsilon$ alone. Denote this process by Y^ε. In finite dimensions Y^ε is directly seen to deviate after a deterministic uniform relaxation time s_{r^*} to a large ball $B_{r^*}(0)$ only negligibly from the deterministic solution u uniformly on time intervals of the order of its inter-jump waiting times $t_{i+1} = T_{i+1} - T_i$. Formally

$$\sup_{x \in D^\pm(\varepsilon^\gamma)\cap B_{r^*}(0)} \sup_{T_i \leqslant t \leqslant T_{i+1}} \|Y^\varepsilon(t) - u(t)\| \to 0 \qquad \text{for} \quad \varepsilon \to 0+ \tag{1.6}$$

in probability. This means that as long as there are no large jumps the solution of the Chafee–Infante equation follows the deterministic solutions on their way to relaxation in the neighborhoods of stable equilibria. Therefore they cannot contribute essentially to exits from their domains of attraction. Exits from these domains will thus be triggered by large jumps. Since in infinite dimensions we solve our equation in a mild sense we establish instead of (1.6) that the small deviation result for Y^ε is implied by

$$\varepsilon\xi^*(t) \to 0, \quad \varepsilon \to 0+, \text{ for } t \geqslant 0.$$

Here $\xi^*(t) = \int_0^t S(t-s)\, \mathrm{d}\xi^\varepsilon(s)$ is the stochastic convolution with respect to ξ^ε (see Sect. 3.3).

Step 3. The inter-jump waiting times of η^ε are all independent and possess exponential laws of parameter β_ε, where

$$\beta_\varepsilon := \nu\left(\frac{1}{\varepsilon^\rho} B_1^c(0)\right) \approx \varepsilon^{\alpha\rho},$$

and ν is the jump measure of L for which we assume that it varies regularly of index α. In accordance with exponential laws, they are therefore expected to be of order $\frac{1}{\varepsilon^\alpha}$. For small ε this quantity is much bigger than the ε-dependent component of the relaxation time $T_{rec} + \kappa\gamma|\ln\varepsilon|$ of the deterministic solution u to $B_{\varepsilon^{2\gamma}}(\phi^\pm)$. We can therefore expect that Y^ε has had enough time to relax to a neighborhood of a stable solution before the next big jump occurs, without leaving the reduced domain in the meantime. This jump therefore originates from a position close to an equilibrium. The effects sketched in (1.4), (1.3) and (1.6) therefore combine, and imply that for small ε exit events start in $B_{\varepsilon^{2\gamma}}(\phi^\pm)$ and are most probably triggered by the large jump part $\varepsilon\eta^\varepsilon$. Hence the first exit time $\tau(\varepsilon)$ from $D^\pm$ is expected to be roughly

$$\tau(\varepsilon) \approx \inf\{T_i = \sum_{j=1}^{i} t_j \mid \phi^\pm + \varepsilon\Delta_{t_i} L \notin D^\pm\}.$$

Step 4. Using the regular variation of the Lévy measure ν of L we obtain for the probability of large jumps big enough to trigger exits

$$\begin{aligned}\mathbb{P}\left(\phi^\pm + \varepsilon\Delta_{t_i} L \notin D^\pm\right) &= \mathbb{P}\Big(\Delta_{t_1} L \in \frac{1}{\varepsilon}\left((D^\pm)^c - \phi^\pm\right)\Big)\\ &= \frac{\nu\left(\frac{1}{\varepsilon}\left((D^\pm)^c - \phi^\pm\right) \cap \frac{1}{\varepsilon^\rho} B_1^c(0)\right)}{\nu\left(\frac{1}{\varepsilon^\rho} B_1^c(0)\right)} \approx \varepsilon^{\alpha(1-\rho)}.\end{aligned}$$

Therefore exits times from reduced domains of attraction of the stable equilibria in the limit of small noise are given by

$$\begin{aligned}\mathbb{E}\left[\tau(\varepsilon)\right] &\approx \sum_{i=1}^{\infty} \mathbb{E}\left[T_i\right] \mathbb{P}\left(\inf\{j : \phi^\pm + \varepsilon\Delta_{t_j} L \notin D^\pm\} = i\right)\\ &\approx \mathbb{E}\left[t_1\right] \mathbb{P}\left(\phi^\pm + \varepsilon\Delta_{t_1} L \notin D^\pm\right) \sum_{i=1}^{\infty} i \ \left(1 - \mathbb{P}\left(\phi^\pm + \varepsilon\Delta_{t_1} L \notin D^\pm\right)\right)^{i-1}\\ &\approx \frac{1}{\varepsilon^{\alpha\rho}}\, \varepsilon^{\alpha(1-\rho)} \left(\frac{1}{\varepsilon^{\alpha(1-\rho)}}\right)^2 = \frac{1}{\varepsilon^\alpha}.\end{aligned}$$

1.3 A Glance at Related Literature

To the best of our knowledge the method of this work sketched in Sect. 1.2 has not been used in the context of SPDEs so far. We shall therefore only give an overview over parts of the literature to which our attention had been drawn in the course of these studies. We do not claim completeness.

The Chafee–Infante equation has been extensively studied, starting with the article by [CI74]. Its most interesting feature is a bifurcation in the system parameter representing the steepness of the potential. This considerably changes the dynamics in comparison to the finite dimensional case, see for example [CP89]. Other classical references are the books by [Hen83] and Hale [Hal83]. Existence and regularity of its solutions have been investigated, as well as the fine structure of the attractor. We refer to the books [Tem92, CH98, Rob01, Chu02] and references therein.

SPDE with Gaussian noise go back to the seventies with early works by the authors of [Par75, KR07] and [Wal81, Fre85, Wal86]. Since then the field has expanded enormously in depth and variety, as is impressively documented recently for example in [KRAD$^+$08]. More recent treatments can be found among others for instance in the books and articles [DZ92, Cho07, PR07, Kot08, CF11, CFO11, Hai11, HRW12, Hai13] and references therein.

The treatment of the asymptotic dynamical behavior for finite dimensional Gaussian diffusions mainly by techniques related to large deviations was developed in [FV70, FV98]. In [FJL82], the authors use methods based on large deviations in order to analyze the stochastic dynamics for SPDE with Gaussian noise. The *tunneling effects* they discover interpret the phenomenon of metastable behavior of solutions switching between stable equilibria at time scales exponential in the noise intensity. Additionally they show that the transitions asymptotically take place at the saddle points, the number of which varies according to the bifurcation scenarios of the deterministic part. Martinelli et al. [MOS89] show that suitably normalized exit times are asymptotically exponential. Brassesco [Bra91] shows that the process is asymptotically concentrated in balls around the stable states and that the average along trajectories remains close to the stable state before the switching time.

SPDEs with jump noise have been studied since the late eighties, see for example [CM87] and [KPA88]. At the end of the nineties the subject is taken up again in a rich and ongoing series of articles for example by the authors of [AWZ98, Mue98, Bie98, AW00, FR00, Fou00, Fou01, Myt02, Kno04, Sto05, Hau05, Hau06, BW06, PZ06, RZ07, MPR10, FTT10a, FTT10b, DX10, Pré10, Xie10, Wu10, PZ10, PXZ11]. We refer to the monograph [PZ07] for a comprising view on SPDEs with Lévy noise and the bibliography therein.

1.4 Organization of the Book

The material in this book is organized as follows.

In Chap. 2 we study properties of the solution of the deterministic Chafee–Infante equation (1.1). Some of them, which are useful for our purposes and well-known in the literature for a long time are collected in Sect. 2.1. Among them are for instance the uniform absorption of a large ball by the global attractor in H, as well as its precise complex geometric structure. The subsequent Sect. 2.2 is dedicated to the construction of forward invariant subdomains of attraction with respect to the solution flow, appropriately reduced in several steps with respect to a parameter ε.

In fact, the aim is to retain a fortiori forward invariance for these reduced domains of attraction with respect to ε-dependent tubes around trajectories of the deterministic solution.

The remainder of the section combines several concepts in order to prove *Proposition 2.12*, the main result of the chapter. It states that there are constants $T_{rec}, \kappa, \varepsilon_0 > 0$ such that for all $\gamma \in (0,1)$, $0 < \varepsilon \leqslant \varepsilon_0$, the deterministic solution $u(t,\zeta;x) = X^0(t,\zeta;x)$ starting from x in a reduced domain $D^{\pm}(\varepsilon^{\gamma})$ is absorbed by the open ball $B_{\varepsilon^{2\gamma}}(\phi^{\pm})$ centered in a stable fixed point after time $T_{rec} + \kappa\gamma|\ln\varepsilon|$. Formally

$$u(t;x) \in B_{\varepsilon^{\gamma}}(\phi^{\pm}) \qquad \forall\, t \geqslant T_{rec} + \kappa\gamma|\ln\varepsilon|, \quad x \in D^{\pm}(\varepsilon^{\gamma}).$$

This is actually a forward analogue to the absorption result in finite dimension. But since in infinite dimensions the attractor contains generically heteroclinic connections between unstable states of the system, the question of the exit from neighborhoods of unstable states in the separating manifold has to be carefully treated. In particular for the linearization of the system in the vicinity of unstable points the Hartman–Grobman result is not appropriate due to the lack of smoothness of the conjugation maps. Instead we construct the stable and unstable manifolds and exploit their transversality in order to prove exponential repulsion from unstable states sitting on the separating manifold in Sect. 2.2.4.

In Chap. 3 we collect some basic and more advanced material about stochastic equations in infinite dimensions, with a particular view towards solutions X^{ε} of the stochastic Chafee–Infante equation. We introduce Lévy processes with values in Hilbert spaces, and discuss their decomposition into appropriate compound Poisson large jump components and small jump components. We give a brief introduction to the theory of stochastic integration for Lévy processes, and of global existence and uniqueness of solutions X^{ε} with respect to the concept of mild solutions. This is discussed along with stochastic convolutions with Lévy noise. The chapter ends with a discussion of the strong Markov property and its consequences in the particular case of the stochastic Chafee–Infante equation, and the presentation of basic material on slowly and regularly varying functions. These concepts are needed in the context of the jump measures of the driving Lévy processes arising in our stochastic equations.

Chapter 4 is devoted to the derivation of the crucial small deviation result of the solution of the Chafee–Infante equation perturbed only by the small jump part of the driving Lévy process from the solution of the deterministic equation. It is here that the technique of decomposition of the Lévy process into a small and large jump component starts taking effect. Assume for simplicity that the Lévy process L is a pure jump process with symmetric Lévy measure ν, which is regularly varying of index $\alpha \in (0,2)$. Then $L = \xi^{\varepsilon} + \eta^{\varepsilon}$ can be decomposed into the martingale ξ^{ε} with jumps bounded from above $\|\Delta_{\cdot} L\| \leqslant \frac{1}{\varepsilon^{\rho}}$, $\rho \in (0,1)$, and the compound Poisson process η^{ε} with finite intensity $\beta_{\varepsilon} = \nu\left(\frac{1}{\varepsilon^{\rho}} B_1^c(0)\right)$ and the jump measure

$\nu\left(\cdot \cap \frac{1}{\varepsilon^\rho} B_1^c(0)\right)/\beta_\varepsilon$. By regular variation of ν, the rate β_ε turns out to be of the order $\varepsilon^{\alpha\rho}$ for small ε.

For $n \in \mathbb{N}$ let T_n be the n-th jump of η^ε. Then due to the structure of the mild solution X^ε the increments $X^\varepsilon(T_1;x) - X^\varepsilon(T_1-;x)$ and $\varepsilon(L(T_1) - L(T_1-))$ coincide. By the strong Markov property it follows for $t \leqslant T_1$ that $X^\varepsilon(t;x) = Y^\varepsilon(t;x)$, if $Y^\varepsilon(\cdot;x)$ is the solution of (1.2), where L is replaced by the small jump martingale ξ^ε. Since $\varepsilon\xi^\varepsilon$ is of pure jump type for $t \leqslant T_1$ the jump increments $\|X^\varepsilon(t) - X^\varepsilon(t-)\| = \|Y^\varepsilon(t) - Y^\varepsilon(t-)\|$ equal $\varepsilon\|\xi^\varepsilon(t) - \xi^\varepsilon(t-)\|$ and hence are bounded by $\varepsilon^{1-\rho} \searrow 0$ as $\varepsilon \to 0+$. It is therefore reasonable to expect the convergence $Y^\varepsilon(t;x) \to u(t;x)$ in an appropriate sense as $\varepsilon \to 0+$. In fact in *Proposition 4.7* this turns out to true for fixed time horizon T and initial values x in a bounded subset of $D^\pm(\varepsilon^\varepsilon)$. In order to ensure the mentioned boundedness condition on the initial values we prove in Sect. 4.1 with the help of perturbation arguments that in the presence of bounded noise $\|\varepsilon\xi^*\| \leqslant 1$ the small noise solution Y^ε enters a ball $B_{r^*}(0)$ before a deterministic time $s_{r^*} > 0$.

Eventually, proceeding from deterministic to random time intervals T_1 in Sects. 4.2 and 4.3 we prove in the crucial *Proposition 4.5* that there are right choices of ρ, γ providing a constant $\vartheta > \alpha(1-\rho)$ such that the *small deviations event*

$$E_x := \{ \sup_{s\in[0,s_{r^*}]} \|\varepsilon\xi^*\| \leqslant \varepsilon^{2\gamma}, \quad \sup_{s\in[s_{r^*},T_1]} \|Y^\varepsilon(s;x) - u(s-s_{r^*};Y^\varepsilon(s_{r^*};x))\| \leqslant (1/2)\varepsilon^{2\gamma} \},$$

has small probability uniformly in the initial position x. More precisely there exists $C_\theta > 0$ and ε_0 such that for $0 < \varepsilon \leqslant \varepsilon_0$

$$\mathbb{P}(\cup_{x\in D^\pm(\varepsilon^\gamma)} E_x^c) \leqslant C_\vartheta \varepsilon^\vartheta.$$

Chapter 5 starts with estimates of probabilities for exit events of X^ε by those of events of the form $\{T_1 > s_{r^*} + T_{rec} + \kappa|\ln\varepsilon|\}$, E_x and $\{\phi^\pm + \varepsilon W_1 \in D^\pm(\varepsilon^\gamma)\}$, where $W_1 = X^\varepsilon(T_1) - X^\varepsilon(T_1-)$ is the size of the first big jump. Under some mild non-degeneracy conditions on the Lévy characteristics of our noise process, we are able to prove the main *Theorem 5.11* about exponential convergence of first exit times of the reduced domains of attraction $D^\pm(\varepsilon^\gamma)$. This is done in a sequence of theorems along the lines of arguments explained in Sect. 1.2, and via a calculation of Laplace transforms of exit times in the small noise limit. We eventually construct a family of random variables $\bar{\tau}(\varepsilon)_{\varepsilon>0}$ with $\mathscr{L}(\bar{\tau}(\varepsilon)) = EXP(1)$ for all $\varepsilon > 0$ such that for all $\theta < 1$

$$\lim_{\varepsilon\to 0+} \mathbb{E}\left[\exp\left(\theta\lambda^\pm(\varepsilon)\tau^\pm(\varepsilon)\right) - \exp\left(\theta\bar{\tau}(\varepsilon)\right)\right] = 0.$$

In Chap. 6 exit times are used to investigate the asymptotic behavior of transition times between different domains of attraction of the Chafee–Infante equation. We first apply the results obtained before to estimate entering times into different reduced domains of attraction (*Theorem 6.3*). This again leads to the description of the asymptotic behavior of the transition times between small balls around different stable equilibrium states in the small noise limit (*Theorem 6.7*).

Chapter 7 starts with a detailed discussion of an additional hypothesis on the jump characteristics of the driving Lévy process, which provides an upper bound for the time to leave neighborhoods of the separating manifold between the domains of attraction.

In Sect. 7.2 we derive two localization results for the solution of the stochastic Chafee–Infante equation on subcritical and critical time scales. Section 7.3 is devoted to the main result of this work, the description of the metastable behavior of the stochastic Chafee–Infante equation (*Theorem 7.10*). It states the convergence of the solution of the stochastic Chafee–Infante equation to a continuous time Markov chain switching between the stable states $\phi^{\pm}$ on a critical time scale which corresponds to the typical exit time scale of Chap. 5. The Markov chain's switching rates are directly related to the mass of the centered domains of attraction $D^{\pm} - \phi^{\pm}$ with respect to the limiting measure of the regularly varying Lévy jump measure ν.

The appendix provides a more detailed treatment of some aspects of the climate physics background leading to the study of the dynamics of one-dimensional stochastic differential equations perturbed by Lévy noise. It is derived from energy balance models in [IP08], and—in an idealized version—the dynamics of the Chafee–Infante equation. We briefly review basic ideas of low dimensional models, and explain the heuristics of coupled atmosphere-ocean models investigated by Hasselmann [Has76] which in a scaling limit are believed to provide nonlinear S(P)DE describing qualitative features of climate dynamics. We finally discuss the simple class of noisy energy-balance models which, if Milankovich cycles as a source of periodic forcing are fed into the system, lead to a qualitatively correct explanation of the dynamics of global glacial periods.

Chapter 2
The Fine Dynamics of the Chafee–Infante Equation

In this chapter, we introduce the deterministic Chafee–Infante equation. This equation has been the subject of intense research and is very well understood now. We recall some properties of its longtime dynamics and in particular the structure of its attractor. We then define reduced domains of attraction that will be fundamental in our study and give a result describing precisely the time that a solution starting form a reduced domain of attraction needs to reach a stable equilibrium. This result is then proved using the detailed knowledge of the attractor and classical tools such as the stable and unstable manifolds in a neighborhood of an equilibrium.

2.1 The Classical Dynamics of the Chafee–Infante Equation

2.1.1 General Properties of the Solution

For our analysis we work on the following spaces

For $p \geqslant 1$, the norm on the Banach space $L^p(0,1)$ of equivalence classes of functions on the unit interval Lebesgue integrable in the p-th power will be denoted by $|\cdot|_p$. In the case of the Hilbert space $L^2(0,1)$, we drop the subscript and simply write $|\cdot|$, and denote the corresponding scalar product by $\langle\cdot,\cdot\rangle$. Our processes usually will be supposed to take their values on the separable Hilbert space $H = H_0^1(0,1) := \overline{\mathscr{C}_c^\infty(0,1)}^{\|\cdot\|}$, normed by

$$||u|| := \Big(\int_0^1 (\nabla u(\zeta))^2 d\zeta\Big)^{\frac{1}{2}} = |\nabla u| = \langle \nabla u, \nabla u\rangle^{\frac{1}{2}}, \quad u \in H,$$

where ∇u is written for the derivative of $u \in H$ in the sense of generalized functions. We further use the uniform norm for functions usually in the space $C_0(0,1)$ of continuous functions $v \in \mathscr{C}([0,1];\mathbb{R})$ with Dirichlet boundary conditions $v(0) = v(1) = 0$ on the unit interval, and denote it by $|\cdot|_\infty$.

A. Debussche et al., *The Dynamics of Nonlinear Reaction-Diffusion Equations with Small Lévy Noise*, Lecture Notes in Mathematics 2085, DOI 10.1007/978-3-319-00828-8_2,

We denote by $B_r(x)$ the ball in H centered at $x \in H$ with radius r.

The norms can be compared through Poincaré's inequality $|u| \leqslant ||u||, u \in H$, (see e.g. [Bre83]) and $|u|_\infty \leqslant ||u||, u \in H$, which follows from the easiest version of Gauss theorem: For $u \in \mathscr{C}_c^\infty(0,1)$ and $s \in (0,1)$

$$u(s) = u(s) - u(0) = \int_0^s \nabla u(\zeta)\,\mathrm{d}\zeta \leqslant s\,\Big(\int_0^s (\nabla u(\zeta))^2\,\mathrm{d}\zeta\Big)^{\frac{1}{2}} \leqslant \|u\|.$$

Hence we can take the supremum on the left-hand side. The latter just expresses the one-dimensional Sobolev embedding

$$(H, \|\cdot\|) \cong (H_0^1(0,1), |\cdot|_{H_0^1}) \hookrightarrow (\mathscr{C}_0(0,1), |\cdot|_\infty).$$

Unless explicitly stated, all the convergences below are intended in the topology defined by the norm $\|\cdot\|$ of H. We will often use the well-known fact that H is an algebra. This implies for instance that all polynomials are locally Lipschitz on H.

As an example of application we give the following lemma.

Lemma 2.1 (The polynomial nonlinearity is locally Lipschitz in H). *For each $R > 0$ there are $K_{1,R} > 0$ and $K_R > 0$ such that*

$$|f(t) - f(s)| \leqslant K_{1,R}|t-s|, \qquad t, s \in \mathbb{R},\ \textit{with } |t|, |s| \leqslant R, \tag{2.1}$$

$$\|f(u) - f(v)\| \leqslant K_R\|u - v\|, \qquad u, v \in H,\ \textit{with } \|u\|, \|v\| \leqslant R. \tag{2.2}$$

Proof. The proof is found in [SY02], Chap. 5.1.1. We provide it for completeness. The proof of (2.1) is obvious. We show (2.2).

Claim 1. f is locally Lipschitz from $L^2(0,1)$ to H.

We start with $u, v \in B_R(0) \subset H = H_0^1(0,1) \hookrightarrow L^\infty(0,1)$. Due to $|u|_\infty \leqslant |\nabla u|$ for all $u \in H$, we have $|u|_\infty, |v|_\infty \leqslant R$. In particular for each $\zeta \in (0,1)$, $\theta \in [0,1]$

$$|f'(u(\zeta) + \theta(v(\zeta) - u(\zeta)))| \leqslant \sup_{y \in B_R(0)} |f'(y)|_\infty =: K_{1,R} < \infty, \tag{2.3}$$

$$|f''(u(\zeta) + \theta(v(\zeta) - u(\zeta)))| \leqslant \sup_{y \in B_R(0)} |f''(y)|_\infty =: K_{2,R} < \infty.$$

Hence due to the mean value theorem

$$\begin{aligned}
|f(u) - f(v)|_{L^2}^2 &= \int_0^1 (f(u(\zeta)) - f(v(\zeta)))^2\,\mathrm{d}\zeta \\
&= \int_0^1 \Big|\Big(\int_0^1 f'(u(\zeta) + \theta(v(\zeta) - u(\zeta)))\,\mathrm{d}\theta\Big)(u(\zeta) - v(\zeta))\Big|^2\,\mathrm{d}\zeta \\
&\leqslant K_{1,R}^2|u - v|_{L^2}^2 \leqslant K_{1,R}^2\|u - v\|^2.
\end{aligned}$$

Claim 2. f is locally Lipschitz from H to H

For $u, v \in B_R(0) \subset H$ we may calculate

$$\begin{aligned}\|f(u) - f(v)\|^2 &= |f'(u)\nabla u - f'(v)\nabla v|^2_{L^2}\\ &\leqslant 2|f'(u)\nabla u - f'(v)\nabla u|^2_{L^2} + 2|f'(v)\nabla u - f'(v)\nabla v|^2_{L^2}\\ &\leqslant 2|f'(u) - f'(v)|^2\|u\|^2 + 2|f'(v)|^2\|u - v\|^2\\ &\leqslant 2K^2_{2,R}R^2|u - v|^2 + 2K^2_{1,R}\|u - v\|^2 \leqslant \underbrace{2\left((K_{1,R}R)^2 + K^2_{2,R}\right)}_{=:K^2_R}\|u - v\|^2.\end{aligned}$$

□

We next discuss some crucial features of the deterministic system

Existence, uniqueness and regularity results for a large class of deterministic reaction-diffusion equations are well-known for a long time. We summarize them quoting [Tem92], p. 84. The deterministic Chafee–Infante equation is given by

$$\begin{aligned}&\frac{\partial u}{\partial t} - \Delta u + \lambda(u^3 - u) = 0, && t > 0, \zeta \in [0, 1],\\ &u(t, 0) = u(t, 1) = 0, && t > 0,\\ &u(0, \zeta) = x(\zeta), && \zeta \in [0, 1].\end{aligned} \tag{2.4}$$

For its solutions we write $u = (u(t))_{t\geqslant 0}$, or $(u(t; x))_{t\geqslant 0}$ if we wish to emphasize the initial state $x \in H$. For convenience of notation, integrating a function $v \in L^1(0, 1)$ in $\zeta \in [0, 1]$ we often write $\int_0^1 v d\zeta$, omitting the integration parameter ζ.

It is well-known that the solution flow $(t, x) \mapsto u(t; x)$ is continuous in t and x and defines a dynamical system in H. Furthermore the solutions are extremely regular for any positive time, i.e. $u(t) \in \mathscr{C}^\infty(0, 1)$ for $t > 0$.

We use the notation $(S(t))_{t\geqslant 0}$ for the heat semigroup generated by the second derivative with Dirichlet boundary condition. We also set $f(u) = \lambda(u^3 - u)$ so that the mild form of the Chafee–Infante equation has the form:

$$u(t) = S(t)x - \int_0^t S(t - s)f(u(s))ds.$$

It is well-known that $(S(t))_{t\geqslant 0}$ is smoothing and decreases exponentially fast so that we have the following inequalities for $t \geqslant 0$:

$$|S(t)y| \leq e^{-c_0 t}|y|, \quad y \in L^2(0, 1), \quad \|S(t)y\| \leq e^{-c_0 t}\|y\|, \quad y \in H, \tag{2.5}$$

and for $t > 0$:

$$\|S(t)y\| \leq C_1 t^{-1/2} e^{-c_0 t} |y|, \quad |S(t)y|_\infty \leq C_1 t^{-1/4} e^{-c_0 t} |y|, \quad y \in L^2(0,1), \tag{2.6}$$

for some positive constants C_1 and c_0. The following lemma proves uniqueness as well as continuity with respect to the initial data in H for (2.4).

Lemma 2.2. *For any $T > 0$, $R > 0$, there exists a constant $C(T,R)$ such that for all $x, y \in B_R(0)$:*

$$\|u(t;x) - u(t;y)\| \leq C(T,R)\|x - y\|, \quad t \in [0,T].$$

Proof. We set $u(\cdot) = u(\cdot;x)$ and $v(\cdot) = u(\cdot;y)$. Taking the scalar product of (2.4) with $-\Delta u$ and integrating by part gives:

$$\frac{1}{2}\frac{\partial}{\partial t}\|u\|^2 + |\Delta u|^2 + \lambda \int_0^1 3u^2(\nabla u)^2 d\zeta - \lambda\|u\|^2 = 0$$

and

$$\|u(t)\|^2 \leq e^{2\lambda t}\|u(0)\|^2 \leq e^{2\lambda T} R^2.$$

Clearly, the same bound holds for v.

Set $r(t) = u(t) - v(t) = u(t;x) - u(t;y)$, then:

$$\frac{\partial r}{\partial t} - \Delta r + \lambda(u^3 - v^3 - r) = 0.$$

This can be rewritten in the integral form:

$$r(t) = S(t)r(0) + \lambda \int_0^t S(t-s)\left(u^3 - v^3 - r\right) ds.$$

We deduce, since H is an algebra,

$$\begin{aligned} \|r(t)\| &\leq e^{-c_0 t}\|r(0)\| + \lambda \int_0^t e^{-c_0(t-s)} \left\|u^3 - v^3 - r\right\| ds \\ &\leq \|x - y\| + \lambda \int_0^t \left(\frac{3}{2}\left(\|u\|^2 + \|v\|^2\right) + 1\right)\|r\| ds \qquad (2.7) \\ &\leq \|x - y\| + \lambda\left(3e^{2\lambda T}R^2 + 1\right)\int_0^t \|r\| ds. \end{aligned}$$

The result follows then from Gronwall lemma. □

2.1.2 Domains of Attraction and the Global Attractor

Here are the main features of the steady states of the deterministic system

It is known from the classical work [CI74] that the solution u of the Chafee–Infante equation has the following set of fixed points. For a detailed exposition of the bifurcation on the elliptic boundary value problem and the analytic representation of the stationary solutions also consult for instance [Hen83, Hal83, Rau02, Chu02, Wak06] or [Rob01].

Proposition 2.3. *For the Chafee–Infante parameter $\lambda \leqslant \pi^2$ there is a unique stable fixed point $v \equiv 0$. For $\lambda > \pi^2$ there are always two stable fixed points $\phi^\pm \in \mathscr{C}^\infty([0,1])$. More precisely, if $(\pi n)^2 < \lambda \leqslant (\pi(n+1))^2, n \in \mathbb{N}$ there are 2 stable and $(2n-1)$ unstable fixed points $\{0, \phi_j^\mp, j = 1, \ldots, n-2\}$. In other words the set of steady states $\mathscr{E}^\lambda$ has the following shape*

$$\mathscr{E}^\lambda := \begin{cases} \{0\}, & 0 < \lambda \leqslant \pi^2, \\ \{0, \phi^\pm\}, & \pi^2 < \lambda \leqslant (2\pi)^2, \\ \{0, \phi^\pm, \phi_j^\mp, j = 1, \ldots, n-1\}, & (n\pi)^2 < \lambda \leqslant ((n+1)\pi)^2, \qquad n \geqslant 2. \end{cases}$$

Moreover for all $x \in H$ the trajectory $t \mapsto u(t; x)$ converges to an element of $\mathscr{E}^\lambda$. This relies on the fact that there is an energy functional, which may serve as a Lyapunov function for the system. The precise statement is as follows.

Proposition 2.4. *For any $\lambda > 0$ and initial value $x \in H$ there exists a stationary state $v \in \mathscr{E}^\lambda$ of the system (2.4) such that*

$$u(t; x) \to v \text{ in } H \text{ as } t \to \infty.$$

A proof can be found in [CI74, FJL82] and [Hen83]. The next statement characterizes the local properties of the steady states as non-degenerate for generic values of the Chafee–Infante parameter as long as it is large enough.

Proposition 2.5 (Morse–Smale property of fixed points). *For the Chafee–Infante equation with $\lambda \neq (k\pi)^2$ for all $k \in \mathbb{N}$, the fixed points in $\mathscr{E}^\lambda$ are hyperbolic, and the stable and the unstable manifolds of any unstable fixed point $\psi \in \mathscr{E}^\lambda$ intersect transversally.*

A proof is given in [Hen85]. Consult also Sect. 2.2.4.

We fix from now on

- The Chafee–Infante parameter $\pi^2 < \lambda \neq (n\pi)^2$ for all $n \in \mathbb{N}$ in (2.4).

We next summarize properties of the global attractor of the deterministic system

Since our results in Chaps. 4–7 are based on a pathwise analysis we need to further specify the fine structure of the attractor of the Chafee–Infante equation. Note that its shape depends crucially on the bifurcation parameter λ.

The dynamical system induced by the solution flow of Chafee–Infante equation is well-known to have a global attractor $\mathscr{A}^\lambda$ in $L^2(0,1)$, $\mathscr{C}_0(0,1)$ and H. For a proof see [Tem92]. Its shape depends crucially on the parameter λ. Let

$$\begin{aligned}\mathscr{M}^u(v) := \{x \in H \mid \exists\ (u(t))_{t\in\mathbb{R}} &\text{ solution of equation (2.4) in } H \text{ such that}\\ \exists\, t_0 \in \mathbb{R} : x = u(t_0) &\text{ and } \lim_{t\to-\infty} u(t;x) = v\}\end{aligned}$$

be the unstable manifold of $v \in \mathscr{E}^\lambda$. We define for $v, w \in \mathscr{E}^\lambda$ the set of complete connecting orbits

$$\begin{aligned}C(v,w) := \{x \in H \mid \exists\ (u(t))_{t\in\mathbb{R}} &\text{ solution of equation (2.4) in } H \text{ such that}\\ \exists\, t_0 \in \mathbb{R} : x = u(t_0) \text{ and } &\lim_{t\to\infty} u(t;x) = w \text{ and } \lim_{t\to-\infty} u(t;x) = v\}, \quad (2.8)\end{aligned}$$

when it is non-empty. If such an orbit does not exist we set $C(v,w) = \emptyset$. For convenience we introduce the notation: For $v, w \in \mathscr{E}^\lambda$, $v \neq w$

$$v \to w \quad :\Longleftrightarrow \quad C(v,w) \neq \emptyset.$$

Roughly speaking the attractor $\mathscr{A}^\lambda$ consists of all fixed points and all global trajectories $\{u(t), t \in \mathbb{R}\}$. More precisely, following [Chu02], in this case of finitely many fixed points the global attractor has the following shape. For any $\lambda > 0$

$$\mathscr{A}^\lambda = \mathscr{E}^\lambda \cup \bigcup_{v\in\mathscr{E}^\lambda} \mathscr{M}^u(v), \quad \text{where} \quad \mathscr{M}^u(v) = \{v\} \cup \bigcup_{\substack{w\in\mathscr{E}^\lambda\\ v\to w}} C(v,w) \quad (2.9)$$

with the notation established in (2.8). In other words

$$\mathscr{A}^\lambda = \{\phi^+, \phi^-\} \cup \bigcup_{v\in\mathscr{E}^\lambda\setminus\{\phi^+,\phi^-\}} \{v\} \cup \bigcup_{\substack{v,w\in\mathscr{E}^\lambda\\ v\to w}} C(v,w). \quad (2.10)$$

Figure 2.1 sketches the qualitative shape of the attractor for different values of $\lambda > 0$.

Here is a recursive description of the attractor $\mathscr{A}^\lambda$

For $\lambda \in ((n\pi)^2, (\pi(n+1))^2)$ the elements of $\mathscr{E}^\lambda$ as well as the entire set $\mathscr{A}^\lambda$ depend continuously on λ. Thus the topological structure of $\mathscr{A}^\lambda$ remains invariant

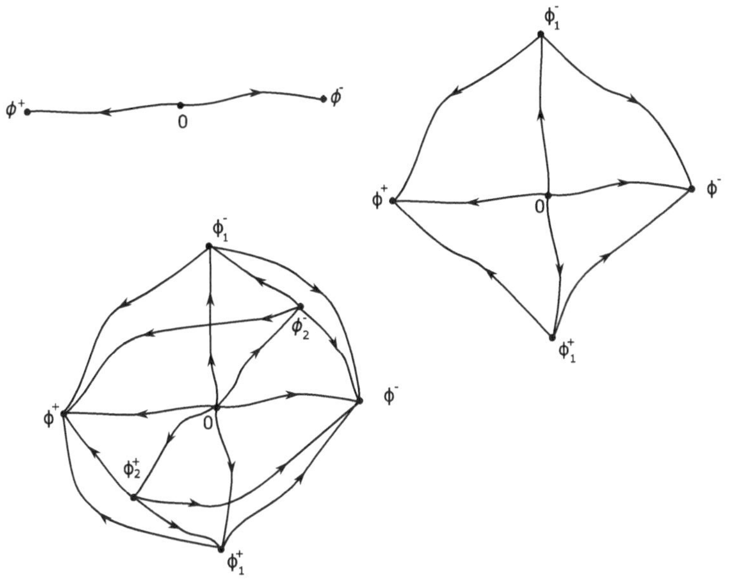

Fig. 2.1 Sketch of $\mathscr{A}^\lambda$ for $\lambda \in (\pi^2, (2\pi)^2)$, $\lambda \in ((2\pi)^2, (3\pi)^2)$, $\lambda \in ((3\pi)^2, (4\pi)^2)$

for $\lambda \in ((n\pi)^2, (\pi(n+1))^2)$. In other words, in this interval all $\mathscr{A}^\lambda$ are homotope. If λ passes $(\pi n)^2$ from the left the connection structure of the elements of $\mathscr{E}^\lambda$ for $\lambda \in ((n\pi)^2, (\pi(n+1))^2)$ is retained in $\mathscr{A}^\lambda$ for $\lambda > (n\pi)^2$ as a substructure, but two new unstable fixed points $\phi_n^\pm$ appear in $\mathscr{E}^\lambda$. In addition, exactly $2n$ new connecting orbits emerge in the attractor: $2(2n-1)$ ones linking the $2n-1$ previously unstable fixed points $\{0, \phi_j^\pm, j = 1, \ldots, n-1\}$ with each of the new ones $\{\phi_n^+, \phi_n^-\}$, and 4 trajectories directed from each the latter ones to each of the stable points $\{\phi^+, \phi^-\}$ and hence $2(2n+1)$ newly connected orbits.

There is some literature on further properties of attractors for reaction diffusion equations, see instance the survey article by Fiedler and Scheel [FS82]. It turns out to be important in the proof of Proposition 2.12 that the longest cascade, i.e. the longest directed sequence of connecting orbits $v_1 \to v_2 \to \cdots \to v_m$ with $v_m \in \{\phi^+, \phi^-\}$, visits $n-1$ fixed points and any cascade ends in one of the stable points $\phi^\pm$. In particular the number of connecting orbits for $\lambda \in ((\pi n)^2, (\pi(n+1))^2)$ is exactly

$$\sum_{k=0}^{n-1} 2(2k+1) = \sum_{k=1}^{n} 2(2k-1) = 4\frac{n(n+1)}{2} - 2n = 2n^2.$$

A special feature of the attractor is that it is uniformly absorbing

All elements $0 \neq v \in \mathscr{E}^\lambda$ are within the unit ball $B_1(0)$ in $\mathscr{C}_0(0,1)$ equipped with $|\cdot|_\infty$ (see [CI74]) and $\mathscr{A}^\lambda$ is known to be there, too [EFNT94]. Since any open set containing the attractor is absorbing for bounded sets, we obtain the following result.

Proposition 2.6. *For any Chafee–Infante parameter $\lambda > 0$ and any $r > 0$ there is a uniform time $T_{rec}^r(\lambda) > 0$, such that for all $t > T_{rec}^r(\lambda)$*

$$\sup_{x \in H} |u(t;x)|_\infty \leqslant 1 + r. \tag{2.11}$$

In [EFNT94], the following result on the relaxation speed

$$\sup_{x \in H} |u(t;x)|_\infty < \sqrt{2}$$

for any $t \geqslant \frac{1}{\lambda}$ is proved. Note that this property implies that the polynomial nonlinearity becomes uniformly Lipschitz in finite time.

Definition 2.7. For $\lambda > \pi^2$ the solution of system (2.4) has two stable stationary solutions, which we shall denote throughout by ϕ^+ and ϕ^-. The full domains of attraction are denoted by

$$D^\pm := \{x \in H \mid \lim_{t \to \infty} u(t;x) = \phi^\pm\},$$

and the separatrix by

$$\mathscr{S} := H \setminus \left(D^+ \cup D^-\right).$$

We use the symbol $\pm$ whenever we can choose simultaneously for all those symbols either $+$ or $-$. In this sense we define the reshifted domains by

$$D_0^\pm := D^\pm - \phi^\pm.$$

Due to the Morse–Smale property the separatrix is a closed $\mathscr{C}^1$-manifold without boundary in H of codimension 1 separating D^+ from D^-. All unstable fixed points lie in $\mathscr{S}$. See [Rau02].

Remark 2.8. In Definition 2.7 as well as in the definition of the unstable manifold above the limits are with respect to the topology of H. However, thanks to the smoothing property of the Chafee–Infante equation it can be shown that the definition does not change if the topology of $L^2(0,1)$ or of $C([0,1])$ is used.

2.2 Logarithmic Bounds on the Relaxation Time in Reduced Domains

2.2.1 Reduced Domains of Attraction

We now define reduced domains of attractions which will be particularly useful to study the dynamic of the Chafee–Infante equation with Levy noise. In the definition below, $D(\mathbb{R}^+; H)$ is the space of càdlàg functions on $\mathbb{R}^+ = [0, \infty)$ with values in H (Fig. 2.2).

Definition 2.9. For $\theta \in L^\infty(0, \infty; H)$ and $x \in H$ we define $v_\theta(\cdot; x)$ by

$$\begin{cases} \dfrac{\mathrm{d}v_\theta}{\mathrm{d}t} = \dfrac{\partial^2 v_\theta}{\partial \zeta^2} + f(v_\theta + \theta) \\ v_\theta(0; x) = x. \end{cases} \tag{2.12}$$

For $\delta_1, \delta_2, \delta_3, \delta_4 \in (0, 1)$, we define the following reduced domains of attractions:

$$D^\pm(\delta_1) := \{x \in D^\pm \mid \cup_{t \geqslant 0} B_{\delta_1}(u(t; x)) \subset D^\pm\} \tag{2.13}$$

$$D^\pm(\delta_1, \delta_2) := \{x \in D^\pm \mid \forall\, \theta \in D(\mathbb{R}^+; H) \text{ with } \sup_{t \geqslant 0} \|\theta(t)\| \leqslant \delta_2 : \\ \cup_{t \geqslant 0} B_{\delta_2}(v_\theta(t; x)) \subset D^\pm(\delta_1)\}, \tag{2.14}$$

$$D^\pm(\delta_1, \delta_2, \delta_3) := \{x \in D^\pm \mid \forall\, \theta \in D(\mathbb{R}^+; H) \text{ with } \sup_{t \geqslant 0} \|\theta(t)\| \leqslant \delta_3 : \\ \cup_{t \geqslant 0} B_{\delta_3}(v_\theta(t; x)) \subset D^\pm(\delta_1, \delta_2)\}, \tag{2.15}$$

$$D^\pm(\delta_1, \delta_2, \delta_3, \delta_4) := \{x \in D^\pm \mid B_{\delta_4}(x) \subset D^\pm(\delta_1, \delta_2, \delta_3)\}. \tag{2.16}$$

For $\gamma, \varepsilon \in (0, 1)$ the sets $\tilde{D}^\pm(\varepsilon^\gamma) := D^\pm(\varepsilon^\gamma, \varepsilon^{2\gamma})$ and $\hat{D}^\pm(\varepsilon^\gamma) := D^\pm(\varepsilon^\gamma, \varepsilon^{2\gamma}, \varepsilon^{2\gamma})$ will be of particular importance to our analysis. Analogously we define the reshifted domains of attraction

$$\begin{aligned} D_0^\pm(\delta_1) &:= D^\pm(\delta_1) - \phi^\pm, \\ D_0^\pm(\delta_1, \delta_2) &:= D^\pm(\delta_1, \delta_2) - \phi^\pm, \\ D_0^\pm(\delta_1, \delta_2, \delta_3) &:= D^\pm(\delta_1, \delta_2, \delta_3) - \phi^\pm, \\ D_0^\pm(\delta_1, \delta_2, \delta_3, \delta_4) &:= D^\pm(\delta_1, \delta_2, \delta_3, \delta_4) - \phi^\pm, \end{aligned}$$

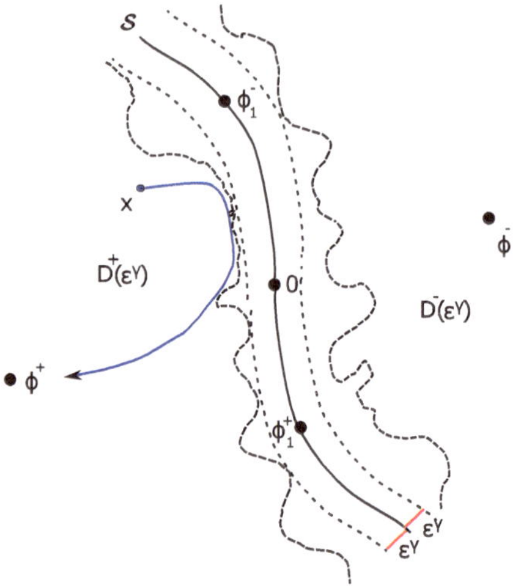

Fig. 2.2 The reduced domains of attraction $D^{\pm}(\varepsilon^{\gamma})$

with the particularly important $\tilde{D}_0^{\pm}(\varepsilon^{\gamma}) = D_0^{\pm}(\varepsilon^{\gamma}, \varepsilon^{2\gamma})$, and $\hat{D}_0^{\pm}(\varepsilon^{\gamma}) = D_0^{\pm}(\varepsilon^{\gamma}, \varepsilon^{2\gamma}, \varepsilon^{2\gamma})$, and the following neighborhood of the separatrix $\mathscr{S} = H \setminus \left(D^+ \cup D^-\right)$

$$\tilde{D}^0(\varepsilon^{\gamma}) := H \setminus \left(\hat{D}^+(\varepsilon^{\gamma}) \cup \hat{D}^-(\varepsilon^{\gamma})\right).$$

For consistency we need the following elementary but non-trivial lemma whose proof is postponed to the end of the chapter.

Lemma 2.10. *For any $\gamma \in (0, 1)$ we have*

1. $D^{\pm} = \bigcup_{\varepsilon>0} D^{\pm}(\varepsilon^{\gamma})$,
2. $D^{\pm} = \bigcup_{\varepsilon>0} \tilde{D}^{\pm}(\varepsilon^{\gamma})$,
3. $D^{\pm} = \bigcup_{\varepsilon>0} \hat{D}^{\pm}(\varepsilon^{\gamma})$.
4. $D^{\pm} = \bigcup_{\varepsilon>0} D^{\pm}(\varepsilon^{\gamma}, \varepsilon^{2\gamma}, \varepsilon^{2\gamma}, \varepsilon^{2\gamma})$.

Moreover, $D^{\pm}(\varepsilon^{\gamma})$, $\tilde{D}^{\pm}(\varepsilon^{\gamma})$, $\hat{D}^{\pm}(\varepsilon^{\gamma})$ and $D^{\pm}(\varepsilon^{\gamma}, \varepsilon^{2\gamma}, \varepsilon^{2\gamma}, \varepsilon^{2\gamma})$ are measurable sets.

A crucial property of $D^{\pm}(\varepsilon^{\gamma})$, $\tilde{D}^{\pm}(\varepsilon^{\gamma})$ and $\hat{D}^{\pm}(\varepsilon^{\gamma})$ is that the deterministic solution starting in one of them will not leave it, that is the positive invariance under the deterministic solution flow.

Lemma 2.11. *The reduced domains of attraction $D^{\pm}(\varepsilon^{\gamma})$, $\tilde{D}^{\pm}(\varepsilon^{\gamma})$ and $\hat{D}^{\pm}(\varepsilon^{\gamma})$ are positively invariant under the deterministic flow, and the following relations are valid with respect to the norm in H:*

(i) $D^{\pm}(\varepsilon^{\gamma}) + B_{\varepsilon^{\gamma}}(0) \subset D^{\pm}$
(ii) $\tilde{D}^{\pm}(\varepsilon^{\gamma}) + B_{\varepsilon^{2\gamma}}(0) \subset D^{\pm}(\varepsilon^{\gamma})$
(iii) $\hat{D}^{\pm}(\varepsilon^{\gamma}) + B_{\varepsilon^{2\gamma}}(0) \subset \tilde{D}^{\pm}(\varepsilon^{\gamma})$
(iv) $D^{\pm}(\varepsilon^{\gamma}, \varepsilon^{2\gamma}, \varepsilon^{2\gamma}, \varepsilon^{2\gamma}) + B_{\varepsilon^{2\gamma}}(0) \subset \hat{D}^{\pm}(\varepsilon^{\gamma})$

Proof. We start with the stated invariances.

1. If $x \in D^{\pm}(\varepsilon^{\gamma})$, by definition $\cup_{t \geq 0} B_{\varepsilon^{\gamma}}(u(t; x)) \subset D^{\pm}$. Hence for $s \geq 0$

$$\cup_{t \geq 0} B_{\varepsilon^{\gamma}}(u(t; u(s; x))) = \cup_{t \geq 0} B_{\varepsilon^{\gamma}}(u(s + t; x)) = \cup_{t \geq s} B_{\varepsilon^{\gamma}}(u(t; x)) \subset D^{\pm}.$$

   This proves that $u(s, x) \in D^{\pm}(\varepsilon^{\gamma})$, hence that $D^{\pm}(\varepsilon^{\gamma})$ is positively invariant.
2. Let us take $x \in \hat{D}^{\pm}(\varepsilon^{\gamma})$ and $s \geq 0$, then for any $\theta \in D(\mathbb{R}^{+}; H)$, $t \geq 0$:

$$v_{\theta}(t; u(s; x)) = v_{\tilde{\theta}_s}(t + s; x)$$

   where $\tilde{\theta}_s(t) = \theta(t - s)$ for $s \leq t$ and $\tilde{\theta}_s(t) = 0$ for $t < s$. Clearly $\tilde{\theta}_s \in D(\mathbb{R}^{+}; H)$ and $\sup_{t \geq 0} \|\tilde{\theta}_s(t)\| \leq \sup_{t \geq 0} \|\theta(t)\|$. Therefore

$$\bigcup_{t \geq 0} B_{\varepsilon^{2\gamma}}(v_{\theta}(t; u(s; x))) = \bigcup_{t \geq 0} B_{\varepsilon^{2\gamma}}(v_{\tilde{\theta}_s}(t + s; x)) \subset \bigcup_{t \geq 0} B_{\varepsilon^{2\gamma}}(v_{\tilde{\theta}_s}(t; x)) \subset D^{\pm}(\varepsilon^{\gamma}).$$

   This proves that $u(s, x) \in \tilde{D}^{\pm}(\varepsilon^{\gamma})$, and therefore the positive invariance of $\tilde{D}^{\pm}(\varepsilon^{\gamma})$.
3. The argument is exactly the same to prove the invariance of $\hat{D}^{\pm}(\varepsilon^{\gamma})$

   We then show the stated inclusions.

(i) If $x \in D^{\pm}(\varepsilon^{\gamma})$ and $y \in B_{\varepsilon^{\gamma}}(0)$, then

$$x + y \in B_{\varepsilon^{\gamma}}(x) = B_{\varepsilon^{\gamma}}(u(0; x)) \subset \cup_{t \geq 0} B_{\varepsilon^{\gamma}}(u(t; x)) \subset D^{\pm},$$

showing 1.

(ii) If $x \in \tilde{D}^{\pm}(\varepsilon^{\gamma})$ and $y \in B_{\varepsilon^{2\gamma}}(0)$, then by definition

$$x + y \in B_{\varepsilon^{2\gamma}}(x) = B_{\varepsilon^{2\gamma}}(v_0(0, x)) \subset \cup_{t \geq 0} B_{\varepsilon^{2\gamma}}(v_0(t; x)) \subset D^{\pm}(\varepsilon^{\gamma}).$$

(iii) If $x \in \hat{D}^{\pm}(\varepsilon^{\gamma})$ and $y \in B_{\varepsilon^{2\gamma}}(0)$, then by definition

$$x + y \in B_{\varepsilon^{2\gamma}}(x) = B_{\varepsilon^{2\gamma}}(v_0(0, x)) \subset \cup_{t \geq 0} B_{\varepsilon^{2\gamma}}(v_0(t; x)) \subset \tilde{D}^{\pm}(\varepsilon^{\gamma}).$$

(iv) This is precisely the definition for $x \in D^{\pm}(\varepsilon^{\gamma}, \varepsilon^{2\gamma}, \varepsilon^{2\gamma}, \varepsilon^{2\gamma})$. □

2.2.2 *Logarithmic Relaxation Times*

The following theorem about the deterministic dynamics on the reduced domain of attraction is fundamental for our purposes.

Proposition 2.12. *Let the Chafee–Infante parameter* $\pi^2 < \lambda \neq (k\pi)^2$ *for* $k \in \mathbb{N}$ *be given. Then there exists an independent finite time* $T_{rec} > 0$ *and a constant* $\kappa > 0$ *such that for each* $\gamma > 0$ *there is* $\varepsilon_0 = \varepsilon_0(\gamma) > 0$*, such that for all* $0 < \varepsilon \leqslant \varepsilon_0$*,* $t \geqslant T_{rec} + \kappa\gamma|\ln\varepsilon|$ *and* $x \in D^{\pm}(\varepsilon^{\gamma})$

$$\|u(t;x) - \phi^{\pm}\| \leqslant (1/2)\varepsilon^{2\gamma}.$$

This means roughly, that as long as the system does not start in an ε^{γ}-neighborhood of the separatrix, it takes a time of only logarithmic order of magnitude in $\varepsilon > 0$ to reach a very small neighborhood of a stable state.

The proof is structured into three parts. In Part I we discuss the absorbtion of the trajectories of the Chafee–Infante equation for any initial value $x \in H$ by a neighborhood of the attractor in finite time. This is followed, in Part II, by a detailed discussion of the local behavior of the system when entering different parts of this neighborhood. In other words, using the flow properties we analyze the behaviour of the solution for initial values taking values in the mentioned neighborhood of the attractor. Here we exploit the well-known shape of the attractor and the hyperbolicity of the fixed points. In Part III we can finally use the gradient structure of the system in order to determine the global behavior by the local information gained in Part II.

We fix $\gamma > 0$.

I. Construction of a well chosen neighborhood of the attractor

For any set $A \subset H$ and $\sigma > 0$, we define the *σ-neighborhood of A* by

$$U_\sigma(A) := \bigcup_{x \in A} B_\sigma(x),$$

where $B_\sigma(x)$ denotes the ball centered at $x \in H$ with radius $\sigma > 0$ in the norm of H.

For two fixed points $v, w \in \mathcal{E}^\lambda$ of the Chafee–Infante equation that are connected in $\mathcal{A}^\lambda$, using the notations of Sect. 2.1.2 this means $C(v,w) \neq \emptyset$ or equivalently $v \to w$, we define for $\eta, \sigma > 0$

$$U_\eta(v,w) := U_\eta(C(v,w))$$

the *η-tube around the heteroclinic orbit* $C(v,w) \subset \mathcal{A}^\lambda$ and by

$$U^-_{\eta,\sigma}(v,w) := U_\eta(C(v,w)) \setminus (B_\sigma(v) \cup B_\sigma(w)),$$

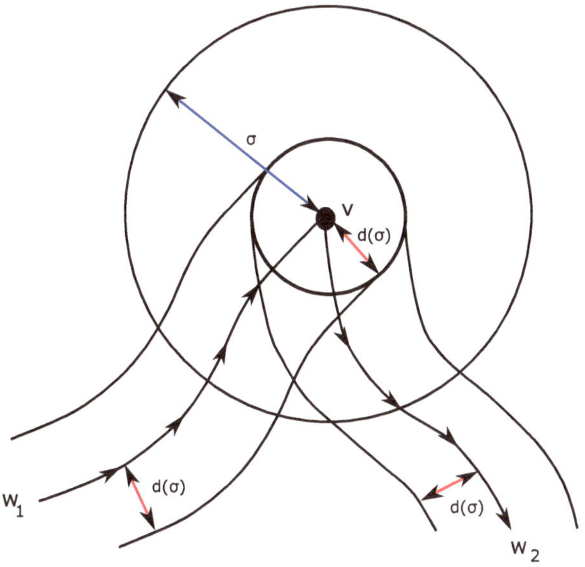

Fig. 2.3 Disjoint tubes around an unstable fixed point $w_1 \to v \to w_2$

the η-tube around the heteroclinic orbit $C(v,w) \subset \mathscr{A}^\lambda$ deprived of the σ-balls around the end points v and w.
For $\sigma \leqslant \frac{\|v-w\|}{3}$ and $\eta > 0$ it follows for all $v, w \in \mathscr{E}^\lambda$ that

$$U^-_{\eta,\sigma}(v,w) \neq \emptyset \qquad \Longleftrightarrow \qquad v \to w. \tag{2.17}$$

For convenience we write $v \leftarrow w$ equivalently to $w \to v$ ($\Leftrightarrow C(v,w) \neq \emptyset$). Define for $\sigma > 0$ the maximal radius $d(\sigma)$ so that the $d(\sigma)$-tubes around the heteroclinic orbits deprived of the σ-balls around the fixed points are all disjoint. More precisely for $\sigma > 0$ we define

$$\begin{aligned} d(\sigma) := \sup\{h > 0 \mid \ & \text{for all } v, w_1, w_2 \in \mathscr{E}^\lambda, \\ & \text{if } w_1 \leftarrow v \to w_2, \text{ then } U_h(v,w_1) \cap U_h(v,w_2) \cap B^c_\sigma(v) = \emptyset, \\ & \text{if } w_1 \to v \leftarrow w_2, \text{ then } U_h(w_1,v) \cap U_h(w_2,v) \cap B^c_\sigma(v) = \emptyset, \\ & \text{if } w_1 \to v \to w_2, \text{ then } U_h(w_1,v) \cap U_h(v,w_2) \cap B^c_\sigma(v) = \emptyset, \\ & \text{if } w_1 \leftarrow v \leftarrow w_2, \text{ then } U_h(v,w_1) \cap U_h(w_2,v) \cap B^c_\sigma(v) = \emptyset\}. \end{aligned} \tag{2.18}$$

For $w_1 \to v \to w_2$, we sketch $d(\sigma)$ in Fig. 2.3.

It is easy to see that due to the transversality of the fixed points there is $\delta_d > 0$ such that for all $0 < \sigma \leqslant \delta_d$ it follows

$$d(\sigma) > 0. \tag{2.19}$$

Note that necessarily $d(\sigma) \leqslant \sigma$.

There exists a constant $\delta_b > 0$ such that for $0 < \sigma \leqslant \delta_b$ the balls $B_\sigma(v), v \in \mathscr{E}^\lambda$, are pairwise disjoint, (2.17) and (2.19) are satisfied, and $B_\sigma(\phi^\pm) \subset D^\pm$. Then by Lemma 2.10 there exists $\varepsilon_b = \varepsilon_b(\sigma) > 0$ such that for $0<\varepsilon\leqslant\varepsilon_b$, $B_\sigma(\phi^\pm) \subset D^\pm(\varepsilon^\gamma)$. We shall exploit the segmented structure (2.10) of attractor $\mathscr{A}^\lambda$ which is reflected in the structure of the surface $\partial U_\sigma(\mathscr{A}^\lambda)$.

For $0 < \sigma \leqslant \delta_b$ and $0 < \eta < d(\sigma)$, we use the following decomposition of $U_{\eta,\sigma}(\mathscr{A}^\lambda)$ of $\mathscr{A}^\lambda$ into disjoint sets

$$U_{\eta,\sigma}(\mathscr{A}^\lambda) = \Big(\bigcup_{v\in\{\phi^+,\phi^-\}} B_\sigma(v) \Big) \cup \Big(\bigcup_{v\in\mathscr{E}^\lambda\setminus\{\phi^+,\phi^-\}} B_\sigma(v) \Big) \cup \Big(\bigcup_{\substack{v,w\in\mathscr{E}^\lambda \\ v\to w}} U^-_{\eta,\sigma}(v,w) \Big). \tag{2.20}$$

Hence for $0 < \varepsilon \leqslant \varepsilon_b$, $0 < \sigma \leqslant \delta_b$ and $0 < \eta \leqslant d(\sigma)$

$$\begin{aligned} &U_{\eta,\sigma}(\mathscr{A}^\lambda) \cap D^\pm(\varepsilon^\gamma) \\ &= B_\sigma(\phi^\pm) \cup \Big(\bigcup_{v\in\mathscr{E}^\lambda\setminus\{\phi^+,\phi^-\}} B_\sigma(v) \cap D^\pm(\varepsilon^\gamma) \Big) \cup \Big(\bigcup_{\substack{v,w\in\mathscr{E}^\lambda \\ v\to w}} U^-_{\eta,\sigma}(v,w) \cap D^\pm(\varepsilon^\gamma) \Big), \end{aligned} \tag{2.21}$$

which by definition of $d(\sigma)$ is a union of pairwise disjoint sets. In the sequel we shall further reduce the upper bounds for σ, η and ε appropriately. We always choose $\eta < d(\sigma)$.

We prove below that all trajectories enter $U_{\eta,\sigma}$ in a bounded time. Then for η, σ sufficiently small, once in this neighbourhood, it has to follow the heteroclinic orbits $C(v,w)$. Since the energy decreases, it can follow only a finite number of such orbits before it enters a neighbourhood of the stable fixed point. In this neighbourhood, the dynamic is very well understood and we conclude our argument.

II. There is a universal relaxation time to enter $U_{\eta,\sigma}(\mathscr{A}^\lambda)$

For any $\sigma > 0$, $d(\sigma) > \eta > 0$ *there is a time* $\tau_1 = \tau_1(\sigma,\eta) > 0$ *such that for all* $t \geqslant \tau_1$ *and* $x \in H$

$$u(t;x) \in U_{\eta,\sigma}(\mathscr{A}^\lambda).$$

Proof. By Temam [Tem92], Remark 1.4, p. 88 there exist $r^* = r^*(\lambda) > 0$ and a uniform upper bound $s_{r^*} = s_{r^*}(\lambda) > 0$ such that for all $t \geqslant s_{r^*}$ and $x \in H$

$$u(t;x) \in B_{r^*}(0). \tag{2.22}$$

By the definition of a global attractor for each $\sigma, \eta > 0$ and each bounded set $A \subset H$ there is a time $t_2 = t_2(A, \sigma, \eta, \lambda) > 0$ such that for all $t \geqslant t_2$ and $x \in A$

$$u(t;x) \in U_{\eta,\sigma}(\mathscr{A}^\lambda)$$

holds true. The claim follows for $A = B_{r^*}(0)$ and $\tau_1 := s_{r^*} + t_2$. □

III. Energy cascades

Consider the energy functional

$$\mathfrak{E}(z) = \int_0^1 \left(\frac{1}{2} \left| \frac{\partial z}{\partial \xi}(\zeta) \right|^2 + \lambda \left(z^4(\zeta) - z^2(\zeta) \right) \right) \mathrm{d}\zeta, \quad z \in H.$$

First note that for all steady states $v, w \in \mathscr{E}^\lambda$, $v \to w$ we have $\mathfrak{E}(v) > \mathfrak{E}(w)$. Indeed, $t \mapsto \mathfrak{E}(u(t;x))$ is non-increasing so that $\mathfrak{E}(v) \geqslant \mathfrak{E}(w)$. Moreover, the equality $\mathfrak{E}(v) = \mathfrak{E}(w)$ would imply that

$$\begin{aligned} 0 &= \frac{\mathrm{d}}{\mathrm{d}t} \mathfrak{E}(u(t;x)) = \nabla \mathfrak{E}(u(t;x)) \frac{\partial u}{\partial t}(t) \\ &= \langle \Delta u(t;x) + f(u(t;x)), \frac{\partial u}{\partial t}(t;x) \rangle = - \left\| \frac{\partial u}{\partial t}(t;x) \right\|^2 \end{aligned}$$

holds for any $x \in C(v,w)$ and $t \in \mathbb{R}$ implying that $C(v,w) \subset \mathscr{E}^\lambda$, which is absurd. Hence for $v \to w$ we have $\mathfrak{E}(v) > \mathfrak{E}(w)$.

We say that $v_1, \ldots, v_k \in \mathscr{E}^\lambda$ form a cascade if $v_1 \to \cdots \to v_k$. Then for any cascade $v_1 \to \cdots \to v_k$ we have

$$\mathfrak{E}(v_1) > \cdots > \mathfrak{E}(v_k).$$

Due to the continuity of $H \ni z \mapsto \mathfrak{E}(z) \in \mathbb{R}$ there is $\delta_g > 0$ such that for $0 < \sigma \leqslant \delta_g$ and each cascade $v_1 \to \cdots \to v_k$ in $\mathscr{E}^\lambda$ we have

$$\sup_{w \in B_\sigma(v_1)} \mathfrak{E}(w) \geqslant \inf_{w \in B_\sigma(v_1)} \mathfrak{E}(w) > \cdots > \sup_{w \in B_\sigma(v_k)} \mathfrak{E}(w) \geqslant \inf_{w \in B_\sigma(v_k)} \mathfrak{E}(w).$$

Obviously, the cascades cannot contain two times the same point of $\mathscr{E}^\lambda$ so that they consist of at most $|\mathscr{E}^\lambda|$ points.

Let $x \in D^{\pm}(\varepsilon^{\gamma})$, we have seen that $u(t,x) \in U_{\eta,\sigma}(\mathscr{A}^{\lambda})$ for $t \geqslant \tau_1(\sigma,\eta)$. Since $D^{\pm}(\varepsilon^{\gamma})$ is invariant, $u(t,x) \in U_{\eta,\sigma}(\mathscr{A}^{\lambda}) \cap D^{\pm}(\varepsilon^{\gamma})$. Thanks to the decomposition (2.21), there are three cases:

- **Case 1:** $u(\tau_1,x) \in B_{\sigma}(v)$ for some $v \in \mathscr{E}^{\lambda} \setminus \{\phi^{\pm}\}$.
- **Case 2:** $u(\tau_1,x) \in U^{-}_{\eta,\sigma}(v,w)$ for some $v,w \in \mathscr{E}^{\lambda}$
- **Case 3:** $u(\tau_1,x) \in B_{\sigma}(\phi^{\pm})$

We define, with the notations of Lemmas 2.13, 2.14 and 2.16,

$$\sigma = \delta_s(\phi^{\pm}) \wedge \min_{v\in\mathscr{E}^{\lambda}\setminus\{\phi^{\pm}\}} \delta_u(v) \wedge \delta_b \wedge \delta_g \wedge \min_{v,w\in\mathscr{E}^{\lambda},v\to w} \delta_h(v,w),$$

$$\varepsilon = \varepsilon_s(\phi^{\pm}) \wedge \min_{v\in\mathscr{E}^{\lambda}} \varepsilon_u(v,\sigma) \wedge \varepsilon_b,$$

$$\eta = \min_{v,w\in\mathscr{E}^{\lambda},v\to w} \eta_2(\sigma,v,w),$$

$$\kappa_0 = \kappa_s(\phi^{\pm}) \vee \max_{v\in\mathscr{E}^{\lambda}\setminus\{\phi^{\pm}\}} \{\kappa_u(v)\},$$

and

$$\bar{\tau} = \max_{v,w\in\mathscr{E}^{\lambda},v\to w} \tau_4(\eta,\sigma,v,w).$$

Case 1: By Lemma 2.14, we know that $u(t,x)$ has to exit $B_{\sigma}(v)$ before the time $\kappa_0\gamma|\ln\varepsilon|$. Since it cannot leave $U_{\eta,\sigma}(\mathscr{A}^{\lambda})$ it has to enter $U^{-}_{\eta,\sigma}(v,w)$ for some $w \in \mathscr{E}^{\lambda}$, so that we are then in Case 2.

Case 2: By Lemma 2.16, it has to enter $B_{\sigma}(w)$ in a time smaller than $\bar{\tau}$ and we are either back to Case 1 or in case 3. Clearly, $\mathscr{E}(v) > \mathscr{E}(w)$.

Thus, the trajectory $u(t,x),\ t \geqslant \tau_1(\sigma,\eta)$ follows the heteroclinic orbits $C(v,w)$ with $v \to w$. Denote by $v_1,\dots,v_k,\dots$ the points in $\mathscr{E}^{\lambda}$ such that it enters $B_{\sigma}(v_i)$. Since $v_i \to v_{i+1}$ they form a cascade. We deduce that there are only a finite number: $v_1 \to \cdots \to v_k$, with $k \leq |\mathscr{E}^{\lambda}|$. Also, all trajectories in $D^{\pm}(\varepsilon^{\gamma})$ enter $B_{\sigma}(\phi^{\pm})$ in a finite time so that necessarily $v_k = \phi^{\pm}$ and we arrive in case 3 in a finite number of step between cases 1 and 2. Each step cannot take more than $\kappa_0\gamma|\ln\varepsilon| + \bar{\tau}$.

In other words, $u(t_0,x) \in B_{\sigma}(\phi^{\pm})$ for some $t_0 \leq \tau_1 + |\mathscr{E}^{\lambda}|\,(\kappa_0\gamma|\ln\varepsilon| + \bar{\tau})$.

IV. Conclusion

In remains to treat the third case.

Case 3: In this case, thanks to Lemma 2.13:

$$\|u(t,x) - \phi^{\pm}\| \leq \frac{1}{2}\varepsilon^{\gamma}$$

for $t \geqslant \tau_1 + \kappa_0\gamma|\ln\varepsilon|$.

It is now easy to gather all possibilities and deduce that

$$\|u(t,x)-\phi^{\pm}\| \leq \frac{1}{2}\varepsilon^{\gamma}$$

for $t \geqslant \tau_1 + |\mathscr{E}^{\lambda}|\,(\kappa_0\gamma|\ln\varepsilon| + \bar{\tau}) + \kappa_0\gamma|\ln\varepsilon|$.

2.2.3 Local Convergence to Stable States

Lemma 2.13 (Local exponential convergence to stable states). *For $\phi \in \{\phi^+,\phi^-\}$ there are constants $\delta_s = \delta_s(\phi) > 0$, $\kappa_s = \kappa_s(\phi) > 0$ and $\varepsilon_s = \varepsilon_s(\phi) > 0$ such that for all $0 < \sigma \leqslant \delta_s$, $0 < \varepsilon < \varepsilon_s$, $y \in B_\sigma(\phi)$ and $t \geqslant \kappa_s\gamma|\ln\varepsilon|$*

$$\|u(t;y)-\phi\| \leqslant (1/2)\varepsilon^{2\gamma}.$$

In addition, there exists $C_2 \geqslant 0$ such that for all $t \geqslant 0$

$$u(t;B_\sigma(\phi)) \subset B_{C_2\sigma}(\phi).$$

Proof. The stability of $\phi \in \{\phi^+,\phi^-\}$ ensures that there exists $\Lambda > 0$ such that for $w \in H$, we have

$$\langle \Delta w + f'(\phi)w, w\rangle \leqslant -\Lambda|w|^2. \tag{2.23}$$

We take $\eta_1 \in (0,1]$ such that for any $w \in B_{\eta_1}(\phi^{\pm})$ we have

$$\|f'(w)-f'(\phi^{\pm})\| \leq \frac{\Lambda}{4}, \quad \|f'(w)\| \leq 2M_1,$$

where $M_1 = \max\{\|f'(\phi^+)\|, \|f'(\phi^-)\|\}$.

For $t \geqslant 0$, $y \in H$ and $\phi \in \{\phi^+,\phi^-\}$ denote by $R(\cdot;y) := u(\cdot;y)-\phi$.

1. We write the equation satisfied by R:

$$\frac{\mathrm{d}R}{\mathrm{d}t} = \Delta R + f'(\phi)R + \left(\int_0^1 \left(f'(\theta\phi+(1-\theta)u) - f'(\phi)\right)\,\mathrm{d}\theta\right)R. \tag{2.24}$$

For convenience we drop the arguments of R. We define for $y \in B_{\eta_1}(\phi)$

$$t^*_{\eta_1} := \inf\{t > 0 \mid \|R(t;y)\| \geqslant \eta_1\}.$$

For $y \in B_{\eta_1}(\phi)$ we have $t^*_{\eta_1} > 0$.

Multiplying (2.24) with R, we obtain for $0 \leqslant t \leqslant t^*_{\eta_1}$

$$\frac{1}{2}\frac{\mathrm{d}}{\mathrm{d}t}|R|^2 + \Lambda |R|^2 \leqslant \frac{\Lambda}{4}|R|^2$$

and obtain,

$$\frac{\mathrm{d}}{\mathrm{d}t}|R|^2 \leqslant -\Lambda |R|^2.$$

Then by Gronwall's Lemma, for $\sigma \leqslant \eta_1$ and $y \in B_\sigma(\phi)$,

$$|R(t)|^2 \leqslant |R(0)|^2 e^{-\Lambda t} \leqslant \sigma^2 e^{-\Lambda t}, \tag{2.25}$$

for $0 \leqslant t \leqslant t^*_{\eta_1}$.

2. We next sharpen the estimate in the first part to an estimate in the $\|\cdot\|$-norm. To this end, we use (2.5) and (2.6):

$$\|S(t)\|_{\mathscr{L}(L^2(0;1);H)} \leq C_1 t^{-1/2} e^{-c_0 t},$$

and

$$\|S(t)\|_{\mathscr{L}(H)} \leq e^{-c_0 t}.$$

We estimate for $t_{\eta_1} \geqslant t \geqslant 0$, $\sigma \leqslant \eta_1$ and $y \in B_\sigma(\phi)$

$$\begin{aligned}\|R(t;y)\| \leqslant & e^{-c_0 t}\|R(0;y)\| + C_1 \int_0^t e^{-c_0(t-s)}(t-s)^{-1/2}|f(u(s;y)) - f(\phi)|\,\mathrm{d}s \\ \leqslant & e^{-c_0 t}\sigma + 2M_1 C_1 \int_0^t e^{-c_0(t-s)}(t-s)^{-1/2}|R(s;y)|\,\mathrm{d}s\end{aligned}$$

Inserting (2.25) we obtain, with $\tilde{\Lambda} = \min\{\Lambda, c_0\}$,

$$\begin{aligned}\|R(t;y)\| \leqslant & e^{-\tilde{\Lambda}t/2}\sigma + 2M_1C_1\left(\int_0^t e^{-c_0(t-s)}(t-s)^{-1/2}\sigma e^{-\tilde{\Lambda}s/2}\,\mathrm{d}s\right) \\ \leqslant & e^{-\tilde{\Lambda}t/2}\sigma + 2M_1C_1\left(\int_0^t e^{-(c_0-\tilde{\Lambda}/2)(t-s)}(t-s)^{1/2}\,\mathrm{d}s\right)\sigma e^{-\tilde{\Lambda}t/2} \\ \leqslant & C_2\sigma e^{-\tilde{\Lambda}t/2}\end{aligned}$$

with $C_2 = 1 + 2M_1C_1\left(\int_0^\infty e^{-(c_0-\tilde{\Lambda}/2)s}s^{-1/2}\,\mathrm{d}s\right)$. We deduce that, if $C_2\sigma \leqslant \eta_1$, then $t^*_{\eta_1} = \infty$ and $u(t,y) \in B_{C_2\sigma}(\phi)$ for all $t \geqslant 0$. Moreover, for $t \geqslant 0$,

$$\|u(t;y) - \phi\| \leqslant C_2\sigma e^{-(1/2)\tilde{\Lambda}t}.$$

Thus we may choose $\kappa_s := \frac{4}{\tilde{\Lambda}}$ and $\delta_s := \min(\frac{\eta_1}{C_2}, \frac{1}{C_2})$. Then for $\sigma \leqslant \eta_1 \leqslant \delta_s$, $\varepsilon_s(\sigma) = \sigma \wedge \exp\big(-\frac{1}{\kappa_s \gamma}\big)$ and $0 < \varepsilon \leqslant \varepsilon_s$, $y \in B_\sigma(\phi)$ and $t \geqslant \kappa_s \gamma |\ln \varepsilon|$

$$\|u(t;y) - \phi\| \leqslant (1/2)\varepsilon^{2\gamma}. \qquad \square$$

2.2.4 Local Repulsion from Unstable States in Reduced Domains

Lemma 2.14. *For given $\lambda > \pi^2$ with $\lambda \neq (\pi n)^2$ for all $n \in \mathbb{N}$ and $v \in \mathscr{E}^\lambda \setminus \{\phi^+, \phi^-\}$ there exists $\delta_u = \delta_u(v) > 0$ and $\kappa_u = \kappa_u(v) > 0$ such that for $0 < \sigma \leqslant \delta_u$ there is $\varepsilon_u = \varepsilon_u(v, \sigma) > 0$ ensuring for $0 < \varepsilon \leqslant \varepsilon_u$ that $D^\pm(\varepsilon^\gamma) \cap B_\sigma(v) \neq \emptyset$ and*

$$\inf\{t \geqslant 0 \,:\, u(t;x) \in B_\sigma^c(v) \cap D^\pm(\varepsilon^\gamma)\} \leq \kappa_u \gamma |\ln \varepsilon|,$$

for $x \in D^\pm(\varepsilon^\gamma) \cap B_\sigma(v)$.

Proof. We prove the result for $v = 0$, the proof clearly extends to any unstable state $v \in \mathscr{E}^\lambda \setminus \{\phi^+, \phi^-\}$.

We are thus interested to determine an upper bound in terms of $\varepsilon > 0$ of the time $u(\cdot\,; x)$ starting in $x \in B_\sigma(0) \cap D^\pm(\varepsilon^\gamma)$ needs to leave $B_\sigma(0) \cap D^\pm(\varepsilon^\gamma)$ for $\sigma > 0, \varepsilon > 0$ sufficiently small. For convenience we recall that the formal Chafee–Infante equation (2.4) for fixed Chafee–Infante parameter $\pi^2 < \lambda \neq (\pi n)^2, n \in \mathbb{N}$, is given as

$$\begin{aligned}
\frac{\partial}{\partial t} u(t,\zeta) &= \Delta u(t,\zeta) + f(u(t,\zeta)), \; \zeta \in [0,1], \; t > 0, \\
u(t,0) &= u(t,1) = 0, \qquad\qquad t > 0, \\
u(0,\zeta) &= x(\zeta), \qquad\qquad\qquad \zeta \in [0,1],
\end{aligned}$$

where $f(y) = -\lambda(y^3 - y)$.

The statement concerns only the dynamic inside $B_\sigma(0)$. Thus, it is sufficient to prove the same result for the Chafee–Infante equation where f is truncated and replaced by

$$f_\sigma(w) = \psi^\sigma(w) f(w), \; w \in H,$$

where the function $\psi^\sigma : H \to H$, $\psi^\sigma \in \mathscr{C}^\infty(H; H)$ is such that

$$\psi^\sigma(w) = \begin{cases} 1 & \text{if } w \in B_\sigma(0) \\ 0 & \text{if } w \in B_{2\sigma}^c(0). \end{cases}$$

In all this proof, we consider only this truncated Chafee–Infante equation whose solution with initial data $x \in H$ is denoted by $u_\sigma(\cdot; x)$.

We set

$$B : D(B) \subset H \to H, \qquad Bw = \Delta + f'(0)w, \qquad w \in D(B)$$
$$G : H \to H, \qquad G(w) = f_\sigma(w) - f'(0)w, \qquad w \in H.$$

Thus $G(0) = G'(0) = 0$. It is well-known that, seen as an unbounded operator on $L^2(0,1)$, B is self-adjoint, has a discrete spectrum, a complete orthonormal system of eigenvectors and a finite number of positive eigenvalues. These eigenvectors are no more orthonormal in H but they still define a complete system of eigenvectors. We denote by $P_u : H \to H^+$ the eigenprojection onto H^+, the span of the eigenvectors of the positive eigenvalues, and respectively $P_s : H \to H^-$, where H^- is the span of the eigenvectors of the negative eigenvalues. Since for the Chafee–Infante parameter $\pi^2 < \lambda \neq (\pi n)^2, n \in \mathbb{N}$, all steady states in $\mathscr{E}^\lambda$ are hyperbolic, 0 is not an eigenvalue and $H = H^+ \oplus H^-$. We denote by $w_s = P_s w$ and $w_u = P_u w$ for $w \in H$. Also, for $t \geqslant 0$ and $x \in D^\pm(\varepsilon^\gamma)$, we use in this subsection the notation

$$Y(t;x) := P_s u_\sigma(t;x),$$
$$Z(t;x) := P_u u_\sigma(t;x).$$

We write

$$g : H^+ \oplus H^- \to H^-, \quad g(w_u, w_s) := P_s G(w),$$
$$h : H^+ \oplus H^- \to H^+, \quad h(w_u, w_s) := P_u G(w)$$

and by $(T(t))_{t\geqslant 0}$ the $\mathscr{C}_0$-semigroup generated by the linearized operator $B = \Delta + f'(0)$. Note that, to lighten notations, we omit to explicit the dependence of g, h and G on σ. The equation satisfied by $u_\sigma(t;x)$ is then equivalent to the coupled system of projected equations

$$Y(t) = T(t)x_s + \int_0^t T(t-\vartheta)g(Y(\vartheta), Z(\vartheta))\, d\vartheta \tag{2.26}$$

$$Z(t) = T(t)x_u + \int_0^t T(t-\vartheta)h(Y(\vartheta), Z(\vartheta))\, d\vartheta \tag{2.27}$$

for $(Y(t), Z(t)) = (Y(t; x_s, x_u), Z(t; x_s, x_u))$.

Clearly, there exists $\omega_0 > 0$ such that for initial values $x_s \in H^-$ we have for all $t \geqslant 0$

$$|T(t)x_s| \leqslant e^{-\omega_0 t}|x_s|.$$

Moreover $T(t)P_u$ can be extended for $x \in H^+$ to $t \leqslant 0$ with

$$|T(t)x_u| \leqslant e^{\omega_0 t}|x_u|.$$

Since, for k large enough, $|(k+B)^{1/2}\cdot|$ defines a norm equivalent to the H norm, it follows that there exists constants c, C such that for $x_s \in H^-$ and $t \geqslant 0$

$$\begin{aligned}\|T(t)x_s\| &\leqslant c|(k+B^{1/2})T(t)x_s| = c|T(t)(k+B^{1/2})x_s| \\ &\leqslant c\,e^{-\omega_0 t}|(k+B^{1/2})x_s| \leqslant cCe^{\omega_0 t}\|x_s\|.\end{aligned}$$

Thus there exists κ_0 such that

$$\|T(t)x_s\| \leqslant \kappa_0 e^{-\omega_0 t}\|x_s\|,\ t \geqslant 0,\ x_s \in H^+.$$

Similarly:

$$\|T(t)x_u\| \leqslant \kappa_0 e^{\omega_0 t}\|x_u\|,\ t \leqslant 0,\ x_u \in H^-.$$

Note that P_s and P_u are not orthogonal but they are bounded operators and we can assume that κ_0 is chosen large enough so that:

$$\|P_s x\| \leq \kappa_0\|x\|, \quad \|P_s x\| \leq \kappa_0\|x\|, \quad x \in H. \tag{2.28}$$

The idea of the proof is that from (2.26) to (2.27), we see that close to the fixed point 0, the dynamic is a perturbation of the linear dynamic governed by $T(t)$. If a point is in $D^{\pm}(\varepsilon^{\gamma})$, it has a non zero projection on H^+. The above inequality may be rewritten as

$$e^{\omega_0 t}\|x_u\| \leqslant \kappa_0\|T(t)x_u\|,\ t \geqslant 0.$$

Therefore a linear orbit escape a neighborhood of 0 exponentially fast. We will see that this is still true for the nonlinear dynamic. We use the stable and unstable manifolds around the fixed point 0.

Definition 2.15. Let $(\Psi(t))_{t\geqslant 0}$ be a dynamical system on H, i.e. a family of continuous operators $\Psi(t) : H \to H$ satisfying

$$\Psi(t+s) = \Psi(t)\circ\Psi(s), \qquad t, s \geqslant 0.$$

Let $v \in H$ a fixed point, i.e. $\Psi(t)v = v$. The *unstable manifold* $\mathscr{W}^u(v)$ *of* v is defined as

$$\mathscr{W}^u(v) := \{w \in H \mid \lim_{t\to-\infty}\Psi(t;w) = v\}.$$

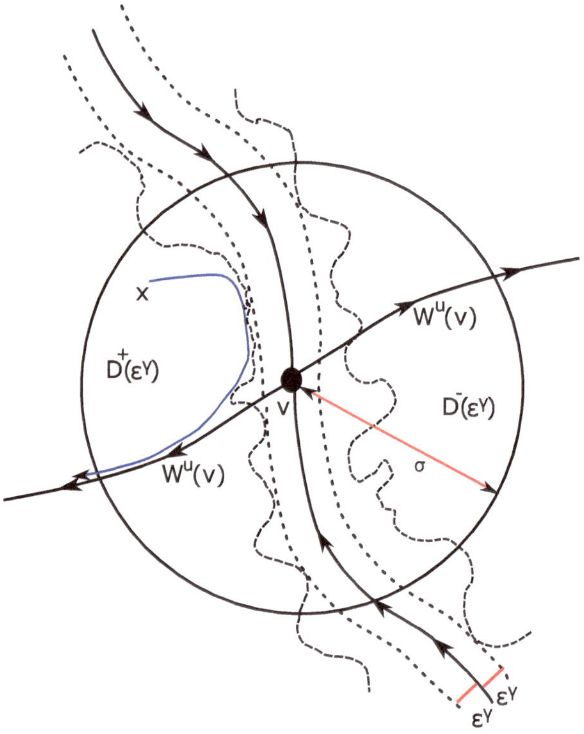

Fig. 2.4 Sketch of the exit from a neighborhood an unstable point

the *stable manifold* $\mathscr{W}^s(v)$ *of* v by

$$\mathscr{W}^s(v) := \{w \in H \mid \lim_{t\to\infty} \Psi(t;w) = v\}.$$

We now examine the structure of the unstable manifold associated to the truncated Chafee–Infante equation (Fig. 2.4).

We denote by $L_\sigma > 0$ the common Lipschitz constant of g and h. Clearly

$$L_\sigma \to 0, \text{ if } \sigma \to 0.$$

In particular, we take σ sufficiently small so that $L_\sigma \leq 1$.

Using the classical Lyapunov–Perron method (see e.g. [Tem92], Chap. IX), it is possible to prove that there exists $\bar{\delta} > 0$ and $\Phi_u : H^+ \to H^-$, a C^1 and globally Lipschitz function with Lipschitz constant $L_{\Phi_u} > 0$ such that for $0 < \sigma \lesssim \bar{\delta}$:

$$\mathscr{W}^u(0) = \{w \in H, P_s w = \Phi_u(P_u w)\}. \tag{2.29}$$

In fact, Φ_u is constructed by a fixed point argument in order to satisfy for all $z \in H^+$

$$\Phi_u(z) = \int_{-\infty}^{0} T(-s)g(\Phi_u(\tilde{Z}(s;z)), \tilde{Z}(s;z))\,\mathrm{d}s, \tag{2.30}$$

where

$$\tilde{Z}(s;z) = T(s)z + \int_{0}^{s} T(s-r)h(\Phi_u(\tilde{Z}(r;z)), \tilde{Z}(r;z))\,\mathrm{d}r, \quad s \leqslant 0. \tag{2.31}$$

Similarly, for $\bar{\delta}$ small enough, there exists and $\Phi_s : H^- \to H^+$, a C^1 and globally Lipschitz function with Lipschitz constant $L_{\Phi_s} > 0$ such that for $0 < \sigma \leqslant \bar{\delta}$:

$$\mathscr{W}^s(0) = \{w \in H, P_u w = \Phi_s(P_s w)\}. \tag{2.32}$$

And Φ_s satisfies for all $y \in H^-$:

$$\Phi_s(y) = -\int_{0}^{\infty} T(-s)h(\tilde{Y}(s,y), \Phi_s(\tilde{Y}(s;y)))\,\mathrm{d}s, \tag{2.33}$$

where

$$\tilde{Y}(s;y) = T(s)y + \int_{0}^{s} T(s-r)h(\tilde{Y}(r,y), \Phi_s(\tilde{Y}(r;y)))\,\mathrm{d}r, \quad s \geqslant 0. \tag{2.34}$$

Moreover, it follows from the construction that $L_{\Phi_u} \to 0$, $L_{\Phi_s} \to 0$ when $\sigma \to 0$.

Claim 1. The graph of Φ_u, $\{w \in H, P_s w = \Phi_u(P_u w)\}$, is negatively invariant and is a subset of $\mathscr{W}^u(0)$.

Proof. It can be assumed that for $0 < \sigma \leqslant \bar{\delta}$

$$\kappa_0 L_\sigma(1 + L_{\Phi_u}) \leqslant \frac{\omega_0}{2}. \tag{2.35}$$

Let $z + \Phi(z)$, $z \in H^+$, be a point on the graph of Φ_u. Since the solution of (2.31) satisfies $\tilde{Z}(s;\tilde{Z}(t;z)) = \tilde{Z}(t+s;z)$, we have for $t \in \mathbb{R}$

$$\Phi_u(\tilde{Z}(t;z)) = \int_{-\infty}^{0} T(-s)g(\Phi_u(\tilde{Z}(s;\tilde{Z}(t;z))), \tilde{Z}(s;\tilde{Z}(t;z)))\,\mathrm{d}s$$

and

$$\begin{aligned}\Phi_u(\tilde{Z}(t;z)) &= \int_{-\infty}^{t} T(t-s)g(\Phi_u(\tilde{Z}(s;z)), \tilde{Z}(s;z))\,\mathrm{d}s \\ &= T(t)\Phi_u(z) + \int_{0}^{t} T(t-s)h(\Phi_u(\tilde{Z}(s;z)), \tilde{Z}(s;z))\,\mathrm{d}s.\end{aligned}$$

Hence for $t \in \mathbb{R}$ and $z \in H^+$ the function $R(t;z) = \Phi_u(\tilde{Z}(t;z)) + \tilde{Z}(t;z)$ satisfies

$$R(t;z) = T(t)R(0;z) + \int_0^t T(t-s)G(R(s;z))\,\mathrm{d}s, \qquad t \leqslant 0.$$

So that it is a solution of the truncated Chafee–Infante equation living on the graph of Φ_u. This shows that the graph of Φ_u is negatively invariant.

In addition, by the Lipschitz continuity of Φ_u and h_σ, we obtain for $t \leqslant 0$

$$\|\tilde{Z}(t;z)\| \leqslant \kappa_0\, e^{\omega_0 t}\|z\| + \kappa_0 \int_t^0 e^{\omega_0(t-\vartheta)} L_\sigma(1 + L_{\Phi_u})\|\tilde{Z}(\vartheta;z)\|\,\mathrm{d}\vartheta.$$

By Gronwall's Lemma

$$\|\tilde{Z}(t;z)\| \leqslant \kappa_0\, e^{(\omega_0 - \kappa_0\, L_\sigma(1+L_{\Phi_u}))t}\|z\|$$

and due to the choice of σ in (2.35) we obtain

$$\|\tilde{Z}(t;z)\| \leqslant \kappa_0\, e^{\frac{\omega_0}{2}t}\|z\| \to 0, \qquad t \to -\infty.$$

Hence

$$\|R(t,z)\| \leqslant \kappa_0\,(1 + L_{\Phi_u})e^{\frac{\omega_0}{2}t}\|z\| \to 0, \qquad t \to -\infty.$$

This shows that the graphs of Φ_u is a subset of the unstable manifold. □

Claim 2. For all $0 < \sigma \leqslant \bar{\delta}$, $x \in H^+$ and $t \geqslant 0$

$$\|Y(t;x) - \Phi_u(Z(t;x))\| \leqslant \kappa_0\, e^{-\frac{\omega_0}{2}t}\|x_s - \Phi_u(x_u)\|,$$

so that the graph of Φ_u is exponentially attracting. In addition, the global trajectories $(u_\sigma(t;x))_{t\in\mathbb{R}}$ are of the form $u_\sigma(t;x) = Z(t;x_u) + \Phi_u(Z(t;x_u))$.

Proof. Let $z \in H^+$. We have for $t \in \mathbb{R}$

$$\Phi_u(\tilde{Z}(t;z)) = \int_{-\infty}^t T(t-s)g(\Phi_u(\tilde{Z}(s;z)); \tilde{Z}(s;z))\,\mathrm{d}s.$$

Thus $\Phi_u(\tilde{Z}(t;z))$ is a mild solution on $\mathbb{R}$ of

$$\frac{\mathrm{d}}{\mathrm{d}t}\Phi_u(\tilde{Z}(t;z)) = B\Phi_u(\tilde{Z}(t;z)) + g(\Phi_u(\tilde{Z}(t;z)), \tilde{Z}(t;z)). \qquad (2.36)$$

By the chain rule

$$\begin{aligned}\frac{\mathrm{d}}{\mathrm{d}t}\Phi_u(\tilde{Z}(t;z)) &= (\nabla\Phi_u)(\tilde{Z}(t;z))\frac{\partial}{\partial t}\tilde{Z}(t;z)\\ &= (\nabla\Phi_u)(\tilde{Z}(t;z))(B\tilde{Z}(t;z) + h(\Phi_u(\tilde{Z}(t;z)), \tilde{Z}(t;z))) \end{aligned} \tag{2.37}$$

we obtain at $t = 0$ that

$$(\nabla\Phi_u)(z)(Bz + h(\Phi_u(z), z)) - B\Phi_u(z) - g(\Phi_u(z), z) = 0 \qquad \text{for all } z \in H^+.$$

Let $(Y(t;x); Z(t;x))_{t\geqslant 0}$ be the decomposition of $u_\sigma(t;x)$. In the next calculation we omit the arguments for convenience. By identification of the right-hand side of (2.36) and (2.37) we obtain

$$\begin{aligned}\frac{\mathrm{d}}{\mathrm{d}t}(Y - \Phi_u(Z)) &= BY + g(Y,Z) - \nabla\Phi_u(Z)(BZ + h(Y,Z))\\ &= B(Y - \Phi_u(Z)) + g(Y,Z) - g(\Phi_u(Z), Z)\\ &\quad + \nabla\Phi_u(Z)(h(\Phi_u(Z), Z) - h(Y,Z)).\end{aligned}$$

Therefore $\eta(t;x) = Y(t;x) - \Phi_u(Z(t;x))$ satisfies for $t \geqslant 0$

$$\begin{aligned}\eta(t;x) = T(t)\eta(0;x) &+ \int_0^t T(t-s)\,(g(Y(s;x), Z(s;x)) - g(\Phi_u(Z(s;x)), Z(s;x)))\;\mathrm{d}s\\ &+ \int_0^t T(t-s)\nabla\Phi_u(Z(s;x))\,(h(\Phi_u(Z(s;x)), Z(s;x)) - h(Y(s;x), Z(s;x)))\;\mathrm{d}s.\end{aligned}$$

Thanks to the Lipschitz continuity of Φ_u, g and h and G, we arrive at

$$\begin{aligned}\|\eta(t;x)\| \leqslant \kappa_0\, e^{-\omega_0 t}\,\|\eta(0;x)\| &+ \kappa_0 \int_0^t e^{-\omega_0(t-s)} L_\sigma \|\eta(s;x)\|\;\mathrm{d}s\\ &+ \kappa_0 \int_0^t e^{-\omega_0(t-s)} L_{\Phi_u} L_\sigma \|\eta(s;x)\|\;\mathrm{d}s.\end{aligned}$$

By Gronwall's Lemma we obtain

$$\|\eta(t;x)\| \;\leqslant\; \kappa_0\, e^{-(\omega_0 - \kappa_0\, L_\sigma(1+L_{\Phi_u}))t}\,\|\eta(0;x)\|.$$

Thus for $0 < \sigma \leqslant \delta_l$ such that (2.35) is true and $x \in H$, we have exponential estimate

$$\|\eta(t;x)\| \;\leqslant\; \kappa_0\,\|\eta(0;x)\| e^{-\frac{\omega_0}{2}t} \qquad \text{for } t \geqslant 0.$$

In other words for $x \in H$

$$\|Y(t;x) - \Phi_u(Z(t;x))\| \leqslant \kappa_0 \, e^{-\frac{\omega_0}{2}t} \|Y(0;x) - \Phi_u(Z(0;x))\|, \qquad \text{for } t \geqslant 0. \tag{2.38}$$

Hence the graph of Φ_u is exponentially attracting for $t \to \infty$. If we choose $x \in H$ such that $(Y(t;x), Z(t;x))$ is a solution defined on $\mathbb{R}^-$ which converges to the unstable state 0 for $t \to -\infty$, it is in particular bounded by a constant, $M > 0$, say

$$\|Y(t;x)\| + \|Z(t;x)\| \leqslant M, \qquad t \leqslant 0.$$

For $t_0 \leqslant 0$ and all $t \geqslant t_0$ we have

$$\|Y(t;x) - \Phi_u(Z(t;x))\| \leqslant \kappa_0 \, e^{-\frac{\omega_0}{2}(t-t_0)} \|Y(t_0;x) - \Phi(Z(t_0;x))\| \leqslant \kappa_0 \, M e^{-\frac{\omega_0}{2}(t-t_0)}.$$

Since the left-hand side does not depend on t_0 we can pass to the limit $t_0 \to -\infty$ implying

$$Y(t;x) = \Phi_u(Z(t;x)) \qquad \text{for } t \in \mathbb{R}.$$

Hence a global solution $((Y(t;x), Z(t;x)))_{t \in \mathbb{R}}$ for $x \in H$ lives on the unstable manifold given as the local graph of Φ_u in $B_\sigma(v)$. This finishes the proof of Claim 2.

□

We now prove that close to the unstable point, a solution with not too small projection on H^+ grows exponentially fast.

Claim 3. For all $0 < \sigma \leqslant \bar{\delta}$, $x \in B_\sigma(0)$

$$\left(\|Z(t,x)\| - \kappa_0 \frac{L_\sigma}{4\omega_0} \varepsilon^\gamma \right) e^{\frac{\omega_0}{2}(s-t)} \leq \kappa_0 \, \|Z(s,x)\|$$

for $s \geqslant t \geqslant t_0$ with

$$t_0 = \frac{2}{\omega_0} \left(\ln 16\kappa_0^2 \sigma - \gamma \ln \varepsilon \right).$$

Proof. With our choice of t_0, we know from (2.28) and (2.38)

$$\|Y(t;x) - \Phi_u(Z(t;x))\| \leqslant \kappa_0 \, e^{-\frac{\omega_0}{2}t} \|P_s x - \Phi_u(P_u x)\| \leqslant \frac{1}{8}\varepsilon^\gamma \tag{2.39}$$

for $t \geqslant t_0$. For any $t, s \in \mathbb{R}$, we have

$$\begin{aligned} Z(t;x) = &\ T(t-s)Z(s;x) + \int_s^t T(t-r)h(Y(r;x), Z(r;x))\, \mathrm{d}r \\ &= T(t-s)Z(s;x) + \int_s^t T(t-r)h(\Phi_u(Z(r;x)), Z(r;x))\, \mathrm{d}r \\ &+ \int_s^t T(t-r)\left(h(Y(r;x); Z(r;x)) - h\left(\Phi_u(Z(r;x)), Z(r;x)\right)\right)\, \mathrm{d}r. \end{aligned}$$

Hence we obtain for $t_0 \leqslant t \leqslant s$ with the help of (2.39)

$$\begin{aligned} \|Z(t;x)\| \leqslant &\ \kappa_0\, e^{\omega_0(t-s)}\|Z(s;x)\| + \kappa_0 \int_t^s e^{\omega_0(t-r)}\, L_\sigma(1+L_{\Phi_u})\|Z(r,x)\|\, \mathrm{d}r \\ &+ \kappa_0 \int_t^s e^{\omega_0(t-r)}\, L_\sigma\, \|\Phi_u(Z(r;x)) - Y(r;x)\|\, \mathrm{d}r \\ &\leqslant \kappa_0\, e^{\omega_0(t-s)}\|Z(s;x)\| + \kappa_0\, L_\sigma(1+L_{\Phi_u}) \int_t^s e^{\omega_0(t-r)}\, \|Z(r,x)\|\, \mathrm{d}r \\ &+\ \kappa_0 \frac{L_\sigma \varepsilon^\gamma}{8} \int_t^s e^{\omega_0(t-r)}\, \mathrm{d}r. \end{aligned}$$

Hence Gronwall's Lemma and (2.35) imply the estimate

$$\|Z(t;x)\| \leqslant \kappa_0\, e^{\frac{\omega_0}{2}(t-s)}\|Z(s;x)\| + \kappa_0 \frac{L_\sigma}{4\omega_0}\varepsilon^\gamma \quad \text{for } t_0 \leqslant t \leqslant s. \tag{2.40}$$

The result follows by rearranging this inequality and replacing t by t_0 and s by t.

□

We are now ready to conclude the proof of Lemma 2.14.

By Lemma 2.10, there exists ε_u such that for $0 < \varepsilon \leqslant \varepsilon_u$, $D^\pm(\varepsilon^\gamma) \cap B_\sigma(0) \neq \emptyset$. Let $x \in D^\pm(\varepsilon^\gamma) \cap B_\sigma(v)$. Then, since the stable manifold is a subset of $\mathscr{S}$, we have thanks to the invariance of $D^\pm(\varepsilon^\gamma)$

$$\|Z(t;x) - \Phi_s(Y(t;x))\| \ \geqslant\ \mathrm{dist}\left((Y(t;x), Z(t;x)); \mathscr{S}\right) \geqslant \varepsilon^\gamma,\ t \geqslant 0. \tag{2.41}$$

We then write

$$\begin{aligned} \|Z(t;x) - \Phi_s(Y(t;x))\| \leqslant &\ \|Z(t;x)\| + \|\Phi_s(Y(t;x))\| \leqslant \|Z(t;x)\| + L_{\Phi_s}\|Y(t;x)\| \\ &\leqslant \|Z(t;x)\| + L_{\Phi_s}\|Y(t;x) - \Phi_u(Z(t;x))\| + L_{\Phi_s}\|\Phi_u(Z(t;x))\| \\ &\leqslant\ (1 + L_{\Phi_s}L_{\Phi_u})\|Z(t;x)\| + L_{\Phi_s}\|Y(t;x) - \Phi_u(Z(t;x))\|. \end{aligned} \tag{2.42}$$

So that, using Claim 3, (2.28) and (2.38), for $s \geqslant t \geqslant t_0$:

$$\begin{aligned} \|Z(s;x)\| &\geqslant \left(\|Z(t,x)\| - \kappa_0 \frac{L_\sigma}{4\omega_0}\varepsilon^\gamma\right) e^{\frac{\omega_0}{2}(s-t)} \\ &\geqslant \left((1+L_{\Phi_s}L_{\Phi_u})^{-1}\left(\varepsilon^\gamma - 2\kappa_0 L_{\Phi_s}\sigma e^{-\frac{\omega_0 t}{2}}\right) - \kappa_0 \frac{L_\sigma}{4\omega_0}\varepsilon^\gamma\right) e^{\frac{\omega_0}{2}(s-t)} \end{aligned} \tag{2.43}$$

Set

$$t_1 = \frac{2}{\omega_0}\left[\ln(4\kappa_0 L_{\Phi_s}) + \ln\sigma - \gamma\ln\varepsilon\right]$$

and

$$t_2 = \frac{2}{\omega_0}\left[\ln(4\kappa_0(1 + L_{\Phi_s}L_{\Phi_u})) + \ln\sigma - \gamma\ln\varepsilon\right]$$

then, for $s \geqslant \max\{t_0, t_1\} + t_2$, we obtain with $t = \max\{t_0, t_1\}$:

$$\|Z(s,x)\| \geqslant \kappa_0\sigma.$$

Since

$$\|u_\sigma(s,x)\| \geqslant \kappa_0^{-1}\|Z(s,x)\|,$$

it follows that $u_\sigma(\cdot,x)$ exit $B_\sigma(0)$ before $\max\{t_0,t_1\} + t_2$. This is then also true for $u(\cdot,x)$ and the claim follows. □

2.2.5 *Uniform Exit from Small Tubes around Heteroclinic Orbits*

In this subsection we prove that the exit times from tubes around heteroclinic orbits can be estimated by a constant.

Lemma 2.16 (Uniform exit time from tubes around connecting orbits). *For $v, w \in \mathscr{E}^\lambda$ with $v \to w$ there is $\delta_h = \delta_h(v,w) > 0$ such that for all $0 < \sigma < \delta_h$ there exists $\eta_2 = \eta_2(\sigma, v, w) > 0$ such that for $0 < \eta \leqslant \eta_2$ we obtain $\tau_4 = \tau_4(\eta,\sigma,v,w) > 0$ ensuring for all $y \in U^-_{\eta,\sigma}(v,w)$ that*

$$u(\tau_4; y) \in B_{\sigma/2}(w).$$

Proof. Let $z \in C(v,w)$. Since $C(v,w) = \{u(t;z), t \in \mathbb{R}\}$ and $\mathfrak{E}(v) > \mathfrak{E}(w)$, $u(t;z) \to v$ for $t \to -\infty$. Hence

$$t^- := \inf\{t \in \mathbb{R} \mid \|v - u(t;z)\| = \sigma\} > -\infty.$$

Since $t \mapsto u(t;z)$ is continuous, $\|u(t^-;z) - v\| = \sigma$. Denote by $z^- := u(t^-;z)$. It is uniquely determined and depends only on σ. We also have $C(v,w)=\{u(t;z^-), t\in\mathbb{R}\}$ and $u(t;z^-) \to w$ for $t \to +\infty$. Hence there exists a time $\tau_4 = \tau_4(\sigma,v,w) > 0$ such that for $t \geqslant \tau_4$

$$\|u(t;z^-) - w\| \leqslant \frac{\sigma}{4}.$$

For all $z \in C_\sigma(v,w) := C(v,w) \cap B^c_\sigma(v) \cap B^c_\sigma(w)$, there exits $s \geqslant 0$ such that $z = u(s,z^-)$. Thus for $t \geqslant \tau_4$:

$$\|u(t;z) - w\| \leqslant \|u(t+s;z^-) - w\| \leqslant \frac{\sigma}{4}.$$

The map $x \mapsto u(\tau_4;x)$ is uniformly continuous on bounded sets. This implies that there is $\eta_2 = \eta_2(\sigma,v,w) > 0$ such that for $0 < \eta \leqslant \eta_2$ the inequality $\|y_1 - y_2\| \leqslant \eta$ implies

$$\sup_{s\in[0,\tau_4]} \|u(s,y_1) - u(s,y_2)\| \leqslant \frac{\sigma}{4}.$$

We can choose $\eta_2 \leqslant \sigma$.

Let $\eta_2 = \frac{\sigma}{4C(\tau_4,R+1)} \wedge 1$ where R is such that $C(v,w)$ is in the ball of radius R and $C(\tau_4,R+1)$ is given by Lemma 2.2. Then for $0 < \eta \leqslant \eta_2$ and $y \in U^-_{\eta,\sigma}(v,w)$, dist $(y, C_\sigma(v,w)) \leqslant \eta$ hence there is $z \in C_\sigma(v,w)$ with $\|y - z\| \leqslant \eta$. We then have

$$\|u(\tau_4;y) - w\| \leqslant \|u(\tau_4;y) - u(\tau_4;z)\| + \|u(\tau_4;z) - w\| \leqslant C(\tau_4,R+1)\eta + \frac{\sigma}{4} \leqslant \frac{\sigma}{2}.$$

□

2.3 Proof of Lemma 2.10

We first estimate the difference between solutions of (2.4) and (2.12) on a bounded time interval. It generalizes Lemma 2.2.

Lemma 2.17. *For any $R \geqslant 1$, $M \geqslant 1$ and $T > 0$, there exists a constant $\kappa_1(T,R,M)$ such that for any $\theta \in D(\mathbb{R}^+;H)$ with $\sup_{[0,T]} \|\theta(t)\| \leq M$ and $x, y \in H$ with $\|x\|, \|y\| \leq R$ we have*

$$\|v_\theta(t;y) - u(t;x)\| \leq \kappa_1(T,R,M)\left(\|x-y\| + \sup_{[0,T]} \|\theta(t)\|\right), \quad t \in [0,T].$$

Proof. We lighten notations and set $v = v_\theta(\cdot;y)$, $u = u(\cdot;x)$. Also, we write $\theta_\infty = \sup_{[0,T]} \|\theta(t)\|$.

We have seen in the proof of Lemma 2.2 that

$$\|u(t)\| \leq e^{\lambda T} R, \quad t \geqslant 0.$$

We derive a similar bound for v. We multiply (2.12) by $-\Delta v$ and integrate on $[0, 1]$. We integrate by parts and use Cauchy–Schwarz inequality:

$$\begin{aligned}
&\frac{1}{2}\frac{d}{dt}\|v\|^2 + |\Delta v|^2 + 3\lambda|(v+\theta)\nabla(v+\theta)|^2 \\
&\quad = 3\lambda \int_0^1 |v+\theta|^2 (\nabla(v+\theta))\nabla\theta d\zeta + \lambda \int_0^1 \nabla(v+\theta)\nabla v d\zeta \\
&\quad \leqslant 3\lambda |v+\theta|_\infty |(v+\theta)\nabla(v+\theta)|\, \|\theta\| + \lambda(\|v\|^2 + \|v\|\,\|\theta\|) \\
&\quad \leqslant 3\lambda|(v+\theta)\nabla(v+\theta)|^2 + \frac{3\lambda}{2}\theta_\infty^2\|v\|^2 + \frac{3\lambda}{2}\theta_\infty^4 + \frac{3\lambda}{2}\|v\|^2 + \frac{\lambda}{2}\theta_\infty^2.
\end{aligned}$$

(Recall that $|\cdot|_\infty \leqslant \|\cdot\|$.)

Since $\theta_\infty \leqslant M$, we deduce

$$\frac{d}{dt}\|v\|^2 \leqslant 3\lambda(M^2+1)\|v\|^2 + \lambda(3M^2+1)M^2,$$

and by Gronwall Lemma

$$\|v(t)\|^2 \leqslant e^{3\lambda(M^2+1)T}(R^2+M^2), \quad t \in [0, T].$$

We now define $r = v - u$ and write the integral equation. It satisfies:

$$r(t) = S(t)(y-x) + \lambda \int_0^t S(t-s)\,(f(v+\theta) - f(u))\,ds. \tag{2.44}$$

Then by (2.5):

$$\begin{aligned}
\|r(t)\| &\leq \|x-y\| + \int_0^t \|f(v+\theta) - f(u)\| ds \\
&\leq \|x-y\| + \lambda(3e^{3\lambda(M^2+1)T}(R^2+M^2)+1) \int_0^t \|r\| + \|\theta\| ds.
\end{aligned}$$

It is now easy to conclude with the help of Gronwall Lemma. □

We also need a continuity result on unbounded time interval. This is possible thanks to the strong stability properties of the fixed point $\phi^{\pm}$.

Lemma 2.18. *For any $R \geqslant 1$, there exists $\kappa_2(R)$, $\delta(R) > 0$ such that if $\sup_{t\geqslant 0}\|\theta(t)\| \leqslant \delta(R)$, $\|x\|, \|y\| \leqslant R$, $x \notin \mathscr{S}$ and $\|x-y\| \leqslant \delta(R)$ then*

$$\|u(t,x) - v_\theta(t,y)\| \leqslant \kappa_2(R)\left(\|x-y\| + \sup_{t\geqslant 0}\|\theta(t)\|\right).$$

Proof. Again, we lighten notations and set $v = v_\theta(\cdot; y)$, $u = u(\cdot; x)$.

Let $\Lambda > 0$ be the constant introduced in Lemma 2.13. For $\phi \in \{\phi^+, \phi^-\}$ and $w \in H$, we have

$$\langle \Delta w + f'(\phi)w, w\rangle \leqslant -\Lambda |w|^2. \tag{2.45}$$

We again take $\eta_1 \in (0, 1]$ such that for any $w \in B_{\eta_1}(\phi^\pm)$ we have

$$\|f'(w) - f'(\phi^\pm)\| \leq \frac{\Lambda}{4}, \ \|f'(w)\| \leq 2M_1,$$

where $M_1 = \max\{\|f'(\phi^+)\|, \|f'(\phi^-)\|\}$. Since $x \notin \mathscr{S}$, there exists $\phi \in \{\phi^+, \phi^-\}$ such that $u(t,x) \to \phi$ as $t \to \infty$ and there exists $T \geqslant 0$ such that

$$\|u(t) - \phi\| \leqslant \frac{\eta_1}{4}, \ t \geqslant T.$$

We take $\delta(R) \leqslant \min\{1, \frac{\eta_1}{2}\}$ such that:

$$2\kappa_1(T, R, 1)\delta(R) \leqslant \frac{\eta_1}{8},$$

where κ_1 is given by Lemma 2.17. Then, for $\sup_{t\geqslant 0} \|\theta(t)\| \leqslant \delta(R)$, $\|x\| \leqslant R$, $\|y\| \leqslant R$, $\|x - y\| \leqslant \delta(R)$ and $t \leqslant T$,

$$\|u(t) - v(t)\| \leqslant \frac{\eta_1}{8}.$$

Therefore

$$\|v(T) - \phi\| \leqslant \frac{3\eta_1}{8}.$$

We introduce the stopping time

$$\tau = \inf\{t \geqslant T, \ \|v - \phi\| > \frac{\eta_1}{2}\}.$$

Set $r(t) = u(t) - v(t)$, then

$$\frac{dr}{dt} = \Delta r + f'(\phi)r + f(u) - f(v + \theta) - f'(\phi)r. \tag{2.46}$$

Note that

$$f(u) - f(v + \theta) - f'(\phi)r = \left(\int_0^1 f'(\mu u + (1 - \mu)(v + \theta))d\mu\right)(r + \theta) - f'(\phi)r.$$

Since $|\cdot|_\infty \leq \|\cdot\|$, we have the following bounds for $T \leqslant t \leqslant \tau$:

$$|f'(\mu u + (1-\mu)(v+\theta)) - f'(\phi)|_\infty \leqslant \frac{\Lambda}{4}, \; |f'(\mu u + (1-\mu)(v+\theta))|_\infty \leqslant 2M_1.$$

Multiply (2.46) by r and integrate in space to deduce, thanks to (2.45),

$$\begin{aligned}\frac{1}{2}\frac{d}{dt}|r|^2 + \Lambda|r|^2 &\leqslant \frac{\Lambda}{4}|r|^2 + 2M_1|\theta|\,|r| \\ &\leqslant \frac{\Lambda}{2}|r|^2 + \frac{2M_1^2}{\Lambda}|\theta|^2.\end{aligned}$$

Hence

$$\frac{d}{dt}|r|^2 + \Lambda|r|^2 \leqslant \frac{4M_1^2}{\Lambda}|\theta|^2.$$

It follows by Gronwall Lemma:

$$|r(t)|^2 \leqslant e^{-\Lambda(t-T)}|r(T)|^2 + \frac{4M_1^2}{\Lambda^2}\theta_\infty^2,$$

where as above we have set $\theta_\infty = \sup_{t\geqslant 0}\|\theta(t)\|$. We now get an estimate in H thanks to (2.44), (2.5) and (2.6):

$$\|r(t)\| \leqslant e^{-c_0(t-T)}\|u(T)-v(T)\| + C_1\int_T^t e^{-c_0(t-s)}|t-s|^{-1/2}|f(u)-f(v+\theta)|ds.$$

Using similar argument as above, for $T \leqslant t \leqslant \tau$,

$$|f(u)-f(v+\theta)| \leqslant 2M_1(|r|+|\theta|).$$

We deduce, thanks to Lemma 2.17 and the previous estimate on $|r|$:

$$\begin{aligned}\|r(t)\| \leqslant\; & e^{-c_0(t-T)}\kappa_1(T,R,1)(\|x-y\|+\theta_\infty) \\ & +2C_1M_1\textstyle\int_T^t e^{-c_0(t-s)}|t-s|^{-1/2}\left(e^{-\Lambda(s-T)}|r(T)|^2 + (1+\frac{4M_1^2}{\Lambda^2})\theta_\infty^2\right)ds.\end{aligned}$$

Setting $\tilde{\Lambda} = \min\{\Lambda, \frac{c_0}{2}\}$, we may write

$$\begin{aligned}&\int_T^t e^{-c_0(t-s)}|t-s|^{-1/2}(e^{-\Lambda(s-T)}|r(T)|^2 + (1+\frac{4M_1^2}{\Lambda^2})\theta_\infty^2)ds \\ &\leqslant \Big(\int_0^\infty e^{-\frac{c_0}{2}s}s^{-1/2}ds\Big)\big(e^{-\tilde{\Lambda}(t-T)}\,|r(T)|^2 + (1+\frac{4M_1^2}{\Lambda^2})\theta_\infty\big) \\ &\leqslant \Big(\int_0^\infty e^{-\frac{c_0}{2}s}s^{-1/2}ds\Big)\big(4K_1^2(T,R,1)\delta(R)^2 + (1+\frac{4M_1^2}{\Lambda^2})\delta(R)^2\big).\end{aligned}$$

It follows that provided $\delta(R)$ is sufficiently small, $\|r(t)\| \leqslant \frac{\eta_1}{4}$ for $t \leq \tau$. It follows that $\tau = \infty$ and the above estimate holds on $[T, \infty)$. Using again Lemma 2.17, the result follows. □

We are now ready to prove Lemma 2.10.

Clearly, $D^{\pm}(\varepsilon^{\gamma})$ is measurable since by continuity of $u(t;x)$ in time, we can write

$$\begin{aligned} D^{\pm}(\varepsilon^{\gamma}) &= \{x \in D^{\pm} \mid \cup_{t\in\mathbb{Q}} B_{\varepsilon^{\gamma}}(u(t;x)) \subset D^{\pm}\} \\ &= \cap_{t\in\mathbb{Q}}\{x \in D^{\pm} \mid B_{\varepsilon^{\gamma}}(u(t;x)) \subset D^{\pm}\} \\ &= \cap_{t\in\mathbb{Q}} \cap_{y\in B_{\varepsilon^{\gamma}}\cap A} \{x \in D^{\pm} \mid u(t;x) + y \in D^{\pm}\}, \end{aligned}$$

where A is a dense subset of H. Similarly, since $D(\mathbb{R}^+; H)$ is separable, we can prove that the three other sets are measurable.

Obviously, the fourth statement implies the three others. We therefore prove this one. We use a contradiction argument and assume that there exists $x \in D^{\pm}$ with $x \notin \cup_{\varepsilon>0} D^{\pm}(\varepsilon^{\gamma}, \varepsilon^{2\gamma}, \varepsilon^{2\gamma}, \varepsilon^{2\gamma})$. Then, setting $\varepsilon_n = \frac{1}{n}$, one can find $y_n^1 \in B_{\varepsilon_n^{2\gamma}}(x)$ such that $y_n^1 \notin \hat{D}^{\pm}(\varepsilon_n^{\gamma})$. Thus, there exists $\theta_n^1 \in D(\mathbb{R}^+; H)$ with $\sup_{t\geqslant 0} \|\theta_n^1\| \leqslant \varepsilon_n^{2\gamma}$, $t_n^1 \geqslant 0$ and $y_n^2 \in B_{\varepsilon_n^{2\gamma}}(v_{\theta_n^1}(t_n^1; y_n^1))$, $y_n^2 \notin \tilde{D}^{\pm}(\varepsilon_n^{\gamma})$. We continue the contradiction argument and find $\theta_n^2 \in D(\mathbb{R}^+; H)$ with $\sup_{t\geqslant 0} \|\theta_n^2\| \leqslant \varepsilon_n^{2\gamma}$, $t_n^2 \geqslant 0$, $t_n^3 \geqslant 0$, $y_n^3 \in B_{\varepsilon_n^{2\gamma}}(v_{\theta_n^2}(t_n^2; y_n^2))$, $y_n^4 \in B_{\varepsilon_n^{\gamma}}(u(t_n^3; y_n^3))$, $y_n^4 \notin D^{\pm}$.

We set $R = \sup_{t\geqslant 0} \|u(t;x)\| + 1$.

We remark first that $y_n^1 \to x$ when $n \to \infty$ so that for n large enough $\|y_n^1\| \leqslant R$. By Lemma 2.18, if n is large enough so that $\varepsilon_n \leqslant \delta(R)$:

$$\|u(t_n^1; x) - v_{\theta_n^1}(t_n^1; y_n^1)\| \leqslant \kappa_2(R)(\|x - y_n^1\| + \sup_{t\geqslant 0} \|\theta_n^1(t)\|).$$

Since $y_n^2 \in B_{\varepsilon_n^{2\gamma}}(v_{\theta_n^1}(t_n^1; y_n^1))$, we deduce that $\|y_n^2 - u(t_n^1; x)\| \to 0$. Similarly we prove that $\|y_n^3 - u(t_n^1 + t_n^2; x)\| \to 0$ and $\|y_n^4 - u(t_n^1 + t_n^2 + t_n^3; x)\| \to 0$. This is clearly impossible since $y_n^4 \in (D^{\pm})^c$ and $d((D^{\pm})^c, (u(t;x))_{t\geqslant 0}) > 0$.

Chapter 3
The Stochastic Chafee–Infante Equation

In this preparatory chapter, the tools of stochastic analysis needed for the investigation of the asymptotic behavior of the stochastic Chafee–Infante equation are provided. In the first place, this encompasses a recollection of basic facts about Lévy processes with values in Hilbert spaces. Playing the role of the additive noise processes perturbing the deterministic Chafee–Infante equation in the systems the stochastic dynamics of which will be our main interest, symmetric α-stable Lévy processes are in the focus of our investigation (Sect. 3.1). We even admit a generalization of the concept of α-stable processes on Hilbert spaces, by admitting jump measures with radial tails corresponding to the ones with radial decay of the order $r^{-\alpha}$ as $r \to \infty$, characteristic for α-stable processes. More precisely, the quotient of the respective tails is supposed to be slowly varying, and the jump measures generalizing the ones of α-stable processes in the way just sketched are called regularly varying of order $-\alpha$ (Sect. 3.6). A crucial concept underlying our analysis of exit and transition dynamics of stochastic Chafee–Infante equations will be discussed: the driving Lévy process with regularly varying jump measure will be, given any noise amplitude $\varepsilon > 0$, decomposed into a large jump component consisting of a compound Poisson process η^ε comprising roughly all jumps bigger than $\varepsilon^{-\rho}$, with a parameter $\rho \in (0, 1)$ to be specified, and a complementary small jump component ξ^ε. We recall the basics of stochastic integration with respect to integrators of the type of ξ^ε (Sect. 3.2), defining a martingale due to symmetry of the underlying jump measure. We then exhibit the fundamental results about the existence of solutions of the stochastic Chafee–Infante equation (Sect. 3.4), in the usual mild sense (Sect. 3.3). Among its key properties, we have a closer look at the strong Markov property (Sect. 3.5).

A. Debussche et al., *The Dynamics of Nonlinear Reaction-Diffusion Equations with Small Lévy Noise*, Lecture Notes in Mathematics 2085, DOI 10.1007/978-3-319-00828-8_3,

3.1 Lévy Processes in Hilbert Space and the Noise Decomposition

In this section we give a brief introduction to Lévy processes in a separable Hilbert space $(H, \|\cdot\|, \langle\cdot,\cdot\rangle)$, concentrated on properties which we exploit frequently in later chapters, in particular the Lévy–Khinchine formula and the Lévy–Itô-decomposition. We conclude this section by stating a corollary of the latter claiming that symmetric pure jump Lévy processes can be decomposed into the sum of a martingale, which possesses all moments, and a compound Poisson process.

Definition 3.1. Let $(\Omega, \mathscr{F}, \mathbb{P})$ be a probability space and H a separable Hilbert space. A stochastic process $(L(t))_{t\geqslant 0}$ is a *Lévy process* in H, if it satisfies

1. $L(0) = 0$,
2. for any $n \in \mathbb{N}$ and $0 \leqslant t_1 < t_2 < t_3 < \cdots < t_n$ the vector of increments

$$(L(t_1) - L(t_0), \ldots, L(t_n) - L(t_{n-1}))$$

   is a family of independent random vectors in H,
3. for $0 \leqslant s < t$

$$\mathscr{L}(L(t) - L(s)) = \mathscr{L}(L(t-s)),$$

   where $\mathscr{L}(X)$ denotes the law of a random vector X in H, and
4. it is continuous in probability, i.e. for any $t \geqslant 0$ and $\eta > 0$

$$\lim_{s\to t} \mathbb{P}\left(\|L(t) - L(s)\| > \eta\right) = 0.$$

Remark 3.2. In general neither the marginal nor the incremental distribution of a Lévy process is given explicitly. However at the level of marginals Lévy processes can be identified by their characteristic functions given by the so-called Lévy–Khinchine formula. For a proof see [PZ07], Theorem 4.27.

Theorem 3.3 (Lévy–Khinchine decomposition). *Let $(L(t))_{t\geqslant 0}$ be a Lévy process in a separable Hilbert space H. Then there exist*

- *a vector $a \in H$,*
- *a nonnegative operator of trace class $Q \in L_1^+(H)$, i.e. for an orthonormal basis $(e_i)_{i\in\mathbb{N}}$ of H*

$$\sum_{i=1}^{\infty} \langle Qe_i, e_i\rangle < \infty,$$

- *and a σ-finite measure $\nu : \mathscr{B}(H) \to [0,\infty]$, where $\mathscr{B}(H)$ is the Borel σ-algebra in H, with $\nu(\{0\}) = 0$ satisfying*

$$\int_H \left(1 \wedge \|y\|^2\right) \nu(\mathrm{d}y) < \infty,$$

such that

$$\mathbb{E}\left[e^{i\langle h, L(t)\rangle}\right] = \exp\left(t\,\psi(h)\right), \quad h \in H, \quad t \geq 0, \tag{3.1}$$

and

$$\psi(h) := i\langle h, a\rangle - \frac{1}{2}\langle Qh, h\rangle + \int_H \left(e^{i\langle h, y\rangle} - 1 - i\langle h, y\rangle \mathbf{1}_{\{0<\|y\|\leqslant 1\}}\right) \nu(\mathrm{d}y).$$

The triple of components (Q,ν,a) is called the *characteristic triple* of $(L(t))_{t\geqslant 0}$. The vector a is called the *drift vector* of $(L(t))_{t\geqslant 0}$, Q the *covariance operator for the Wiener part* and ν the *Lévy measure* or *jump measure*.

Remark 3.4. Due to the special form of (3.1) one can easily verify that for each triple (Q,ν,a) as described in the statement there exists a stochastic process $(L(t))_{t\geqslant 0}$ in H which is a Lévy process with this characteristic triple. However for a given Lévy process $(L(t))_{t\geqslant 0}$ the components in formula (3.1) are not unique in general.

Definition 3.5. 1. A Lévy process with characteristic triple $(0,\nu,0)$ is called *pure jump Lévy process.*

2. A *compound Poisson process* $(Y(t))_{t\geqslant 0}$ in H is a stochastic process $(L(t))_{t\geqslant 0}$ with values in H of the following shape. There is a Poisson process $(\pi(t))_{t\geqslant 0}$ with intensity $\lambda > 0$ and an independent sequence of identically distributed random variables $(X_k)_{k\in\mathbb{N}}$ with values in H such that $\mathbb{P}$-almost surely for all $t \geqslant 0$

$$Y(t) = \sum_{k=0}^{\pi(t)} X_k.$$

3. A *compensated compound Poisson process* $(\bar{Y}(t))_{t\geqslant 0}$ in H is a stochastic process of the shape

$$\bar{Y}(t) = Y(t) - t\lambda \int_H y\mu(\mathrm{d}y), \quad t \geqslant 0,$$

where $(Y(t))_{t\geqslant 0}$ is a compound Poisson process with distribution μ of X_1 and moment $\int_H \|y\|\mu(\mathrm{d}y) = \mathbb{E}\|X_1\| < \infty$. In particular it is a martingale with respect to its natural filtration $(\mathscr{F}_t)_{t\geqslant 0}$, where $\mathscr{F}_t = \sigma(\{Y(s), 0 \leqslant s \leqslant t\})$ for $t \geqslant 0$.

Example 3.6. A pure jump Lévy process $(L(t))_{t\geq 0}$ with jump measure $\nu(H) < \infty$ is a compound Poisson process with intensity $\lambda = \nu(H)$ and the jump measure $\mu(B) = \frac{\nu(B)}{\lambda}$ for $B \in \mathscr{B}(H)$, where $X_k = L(t_k) - L(t_k-)$ and $t_k = \inf\{t > 0 \mid \pi(t) = k\}$ for $k \in \mathbb{N}$.

Proposition 3.7 (De Acosta). *A Lévy process $(L(t))_{t\geq 0}$ in a separable Hilbert space H, with Lévy measure ν of bounded support in H has finite moments of all orders. In other words, if there is $r > 0$ such that $\nu(B_r^c(0)) = 0$, then*

$$\mathbb{E}\left[\|L(t)\|^k\right] < \infty \qquad \text{for all } k \geqslant 1,\ t \geqslant 0.$$

For a proof see [PZ07], Theorem 4.4, p. 39.

There exists a decomposition on a path-wise level the components of which correspond to the three components in the characteristic triple of the Lévy–Khinchine decomposition: deterministic drift, Wiener part and pure jump part. In the following we will see that pure jump Lévy processes can be interpreted as the sum of a compound Poisson process and the limit of a sequence of partial sums of independent compensated compound Poisson processes, the intensity of which tends to infinity.

Theorem 3.8 (Lévy–Itô decomposition). *Let $(L(t))_{t\geq 0}$ be a Lévy process in a separable Hilbert space $(H, \|\cdot\|, \langle\cdot,\cdot\rangle)$ and (Q, ν, a) the corresponding characteristic triple from Theorem 3.3. Then there exist*

1. *a Q-Wiener process $(W(t))_{t\geq 0}$ in H, that is $W(t)$ is a centered Lévy process in H satisfying $\mathbb{E}\left[\|W(t)\|^2\right] < \infty$ and*

$$\mathbb{E}\left[\langle W(t), x\rangle\langle W(s), y\rangle\right] = (t \wedge s)\langle Qx, y\rangle \qquad \forall\, x, y \in H, s, t \geqslant 0,$$

2. *for any sequence of positive radii $\mathbb{N} \ni r_n \searrow 0$ and*

$$O_n := \{y \in H \mid r_{n+1} < \|y\| \leqslant r_n\}$$

a sequence of independent compensated compound Poisson processes $(L_n(t))_{t\geq 0}, n \geqslant 0$, in H with jump measures $\nu_n(B) = \nu(B \cap O_n)$ for $B \in \mathscr{B}(H), n \geqslant 1$,

which satisfy $\mathbb{P}$-almost surely for all $t \geqslant 0$

$$L(t) = at + W(t) + \sum_{n=1}^{\infty} \bar{L}_n + L_0(t), \tag{3.2}$$

$$\bar{L}_n(t) = \left(L_n(t) - t\int_H y\nu_n(\mathrm{d}y)\right), n \geqslant 1. \tag{3.3}$$

Furthermore $(W(t))_{t\geq 0}, (L_n(t))_{t\geq 0}, n \geqslant 0$, are independent. The convergence on the right-hand side in (3.2) holds $\mathbb{P}$-almost surely on bounded intervals of $[0, \infty)$.

For a proof see [PZ07], Chap. 4.5. In this work we will focus on a certain class of pure jump processes.

Definition 3.9. 1. A Lévy process $(L(t))_{t\geqslant 0}$ in H is called *symmetric* if its Lévy measure is symmetric in the sense that

$$\nu(-A) = \nu(A) \quad \text{for } A \in \mathscr{B}(H).$$

2. Fix $\alpha \in (0,2)$. An *α-stable process* $(L(t))_{t\geqslant 0}$ in H is a pure jump Lévy process in H where ν has the specific shape

$$\nu(B) = \int_B \nu(\mathrm{d}y) = \int_B \frac{\mathrm{d}r}{r^{1+\alpha}}\sigma(\mathrm{d}s),$$

where $r = \|y\|$ and $s = y/\|y\|$ and $\sigma : \mathscr{B}(\partial B_1(0)) \to [0,\infty)$ is an arbitrary finite Radon measure on the unit sphere of H.

See for example [AG79] for local limit theorems with α-stable limiting laws and their domains of attraction in Banach spaces.

Proposition 3.10. *Let $(M(t))_{t\geqslant 0}$ be an $(\mathscr{F}_t)_{t\geqslant 0}$-martingale in a separable Hilbert space H on a filtered probability space $(\Omega, \mathscr{F}, (\mathscr{F}_t)_{t\geqslant 0}, \mathbb{P})$. Then $(M(t))_{t\geqslant 0}$ has a càdlàg version which is also an $(\mathscr{F}_t)_{t\geqslant 0}$-martingale.*

Remark 3.11. A symmetric pure jump Lévy process $(L(t))_{t\geqslant 0}$ in H with finite expectation $\mathbb{E}\left[\|L(t)\|\right] < \infty$ for all $t \geqslant 0$ is a martingale with respect to the natural filtration $(\mathscr{F}_t)_{t\geqslant 0}$, with

$$\mathscr{F}_t = \sigma(\{L(s) \mid 0 \leqslant s \leqslant t\}), \quad t \geqslant 0.$$

Moreover it is also martingale with respect to the right-continuous completion $(\bar{\mathscr{F}}_t)_{t\geqslant 0}$, with

$$\bar{\mathscr{F}}_t := \bigcap_{s>t} \bar{\mathscr{F}}_s^0, \qquad t \geqslant 0,$$

where $(\bar{\mathscr{F}}_t^0)_{t\geqslant 0}$ is the completion of $(\mathscr{F}_t)_{t\geqslant 0}$ with respect to the null sets of $\mathscr{F}$, i.e. all subsets of measurable sets with probability zero.

With the help of Proposition 3.7 and Theorem 3.8 we obtain the following decomposition which we will frequently use in the sequel.

Theorem 3.12 (Properties of symmetric pure jump Lévy processes). *Let $(L(t))_{t\geqslant 0}$ be a symmetric pure jump Lévy process in a separable Hilbert space H with Lévy measure ν.*

1. Then there exist

a. a càdlàg $(\bar{\mathscr{F}}_t)_{t\geqslant 0}$-martingale $(\xi(t))_{t\geqslant 0}$ with Lévy measure

$$\nu_\xi(B) := \nu(B \cap B_1(0)) \quad \textit{for } B \in \mathscr{B}(H) \textit{ and } 0 \notin \bar{B},$$

which has finite moments of all orders

b. *and a compound Poisson process* $(\eta(t))_{t\geqslant 0}$ *with intensity* $\lambda = \nu(B_1^c(0))$ *and increment distribution* $\mu(B) = \frac{\nu(B\cap B_1^c(0))}{\lambda}$ *for* $B \in \mathscr{B}(H)$

such that ξ *and* η *are independent and* $\mathbb{P}$*-almost surely for all* $t \geqslant 0$

$$L(t) = \xi(t) + \eta(t).$$

2. *There is a positive trace-class operator* $Q_\xi \in L_+^1(H)$ *such that*

$$\langle Q_\xi u, v\rangle = \int_H \langle u, y\rangle\langle v, y\rangle\, \nu_\xi(\mathrm{d}y), \qquad u, v \in H. \tag{3.4}$$

This provides our crucial

We decompose the driving Lévy process in to "small" and "large" jump in a skilled way

Let $(L(t))_{t\geqslant 0}$ be the càdlàg (*continue à droite avec limites à gauche*, or *right continuous and with left limits*) version of a pure jump Lévy process with values in H with a symmetric Lévy measure ν on its Borel σ-algebra $\mathscr{B}(H)$ satisfying

$$\int_H (1 \wedge \|y\|^2)\nu(\mathrm{d}y) < \infty.$$

We denote the jump increment of L at time $t \geqslant 0$ by $\Delta_t L := L(t) - L(t-)$ and decompose the process $L = \eta^\varepsilon + \xi^\varepsilon$ for $\rho \in (0, 1)$ and $\varepsilon > 0$ in the following way. Denote by η^ε the compound Poisson process with intensity

$$\beta_\varepsilon := \nu\left(\frac{1}{\varepsilon^\rho} B_1^c(0)\right)$$

and the jump probability measure as ν outside the ball $\frac{1}{\varepsilon^\rho} B_1(0)$ by

$$\nu\left(\cdot \cap \frac{1}{\varepsilon^\rho} B_1^c(0)\right)/\beta_\varepsilon. \tag{3.5}$$

We further define the complementary process

$$\xi^\varepsilon := L - \eta^\varepsilon. \tag{3.6}$$

The process ξ^ε will be referred to as "small jumps" process, and η^ε as "large jumps" process respectively. Note that for any $\varepsilon > 0$ the processes ξ^ε and η^ε are independent càdlàg Lévy processes with respective Lévy measures $\nu(\cdot \cap B_{\varepsilon^{-\rho}}(0))$ and $\nu(\cdot \cap B_{\varepsilon^{-\rho}}^c(0))$, but in general of very different properties. ξ^ε is a mean zero martingale in H thanks to the symmetry of ν and possesses finite exponential moments.

Since the process η^ε is compound Poisson we can define its jump times. We set recursively

$$T_0 := 0, \qquad T_k := \inf\left\{t > T_{k-1} \,\middle|\, \|\Delta_t L\| > \varepsilon^{-\rho}\right\}, \quad k \geqslant 1,$$

and the periods between successive large jumps of η_t^ε as

$$t_0 = 0, \qquad t_k := T_k - T_{k-1}, \quad k \geqslant 1.$$

These waiting times are exponentially distributed, formally $\mathscr{L}(t_k) = EXP(\beta_\varepsilon)$. We shall denote the k-th large jump by

$$W_0 = 0, \qquad W_k = \Delta_{T_k} L, \quad k \geqslant 1,$$

with the jump distribution (3.5).

3.2 Stochastic Integration in Hilbert Space

In this section we will discuss stochastic integration in a separable Hilbert space H with respect to a Lévy process.

We fix the following convention. Let $(\Omega, \mathscr{F}, \mathbb{P}, (\mathscr{F}_t)_{t\geqslant 0})$ be a filtered probability space, $(L(t))_{t\geqslant 0}$ a symmetric pure jump Lévy process in H and $L(t) = \xi(t)+\eta(t)$, $t \geqslant 0$, a decomposition according to Theorem 3.12. We will first introduce the stochastic integral with respect to the martingale $(\xi(t))_{t\geqslant 0}$.

Definition 3.13. For a time discretization $0 = t_0 < t_1 < \cdots < t_n$, operators $Y_i \in L(H) := L(H; H)$, and events $A_i \in \mathscr{F}_{t_i}$, $1 \leqslant i \leqslant n$, we call

$$Y(t,\omega) := \sum_{i=0}^{n-1} \mathbf{1}_{A_i}(\omega)\, \mathbf{1}_{(t_i,t_{i+1}]}(t)\, Y_i, \qquad \omega \in \Omega, t \geqslant 0, \tag{3.7}$$

a *simple process* in $L(H)$. Denote by $\mathscr{S}(H)$ the space of all simple processes in $L(H)$. Then we define the *stochastic integral* of a simple process $(Y(t))_{t\geqslant 0}$ in $\mathscr{S}(H)$ by

$$\int_0^t Y(s,\omega)\, \mathrm{d}\xi(s,\omega) := \sum_{i=0}^{n-1} \mathbf{1}_{A_i}(\omega)\, Y_i\, (\xi(t_{i+1}\wedge t,\omega) - \xi(t_i \wedge t, \omega)), \quad t \geqslant 0,\ \omega \in \Omega. \tag{3.8}$$

Lemma 3.14 (Itô isometry for simple processes). *For $Y = (Y(t))_{t\geqslant 0} \in \mathscr{S}(H)$ let*

$$I_t^\xi(Y) := \int_0^t Y(s)\, \mathrm{d}\xi(s), \qquad t \geqslant 0.$$

Then

$$\mathbb{E}\left[\|I_t^{\xi}(Y)\|^2\right] = \mathbb{E}\left[\int_0^t |Y(s)Q_{\xi}^{1/2}|_{L_2(H)}^2 \,\mathrm{d}s\right] \qquad \text{for } t \geqslant 0,$$

where $Q_\xi \in L_1^+(H)$ is the covariance operator of ξ given by Theorem 3.12(2). The operator $Q_\xi^{1/2} \in L_2(H)$ is then the unique Hilbert–Schmidt operator such that

$$Q_\xi^{1/2} Q_\xi^{1/2} = Q_\xi.$$

Proof. First of all for an orthonormal basis $(e_l)_{l\in\mathbb{N}}$ of H

$$\begin{aligned}
\mathbb{E}\left[\|\int_0^t Y(s)\,\mathrm{d}\xi(s)\|^2\right] &= \mathbb{E}\left[\left\|\sum_{i=0}^{n-1} \mathbf{1}_{A_i}\, Y_i\left(\xi(t_{i+1}\wedge t) - \xi(t_i\wedge t)\right)\right\|^2\right] \\
&= \mathbb{E}\left[\sum_{i,k=0}^{n-1} \mathbf{1}_{A_i}\mathbf{1}_{A_k}\, \langle Y_i\left(\xi(t_{i+1}\wedge t) - \xi(t_i\wedge t)\right), \right. \\
&\qquad \left. Y_k\left(\xi(t_{k+1}\wedge t) - \xi(t_k\wedge t)\right)\rangle\right] \\
&= \mathbb{E}\left[\sum_{i,k=0}^{n-1}\sum_{l=1}^{\infty} \mathbf{1}_{A_i}\mathbf{1}_{A_k}\, J_{ikl}\right],
\end{aligned}$$

where

$$\begin{aligned}
J_{ikl} &:= \mathbb{E}\left[\langle Y_i\left(\xi(t_{i+1}\wedge t) - \xi(t_i\wedge t)\right), e_l\rangle\langle Y_k\left(\xi(t_{k+1}\wedge t) - \xi(t_k\wedge t)\right), e_l\rangle \mid \mathscr{F}_{t_k\vee t_i}\right] \\
&= \mathbb{E}\left[\langle \xi(t_{i+1}\wedge t) - \xi(t_i\wedge t), Y_i^* e_l\rangle\langle \xi(t_{k+1}\wedge t) - \xi(t_k\wedge t), Y_k^* e_l\rangle \mid \mathscr{F}_{t_k\vee t_i}\right] \\
&= \begin{cases} 0, & i \neq k \\ ((t_{i+1}\wedge t) - (t_i\wedge t))\ \langle Q_\xi Y_i^* e_l, Y_i^* e_l\rangle, & i = k \end{cases}.
\end{aligned}$$

Hence

$$\begin{aligned}
\mathbb{E}\left[\|\int_0^t Y(s)\,\mathrm{d}\xi(s)\|^2\right] &= \sum_{i=0}^{n-1} \mathbb{P}(A_i)\left((t_{i+1}\wedge t) - (t_i\wedge t)\right) \sum_{l=1}^{\infty}\langle Q_\xi Y_i^* e_l, Y_i^* e_l\rangle \\
&= \sum_{i=0}^{n-1} \mathbb{P}(A_i)\left((t_{i+1}\wedge t) - (t_i\wedge t)\right) \sum_{l=1}^{\infty} \|Q_\xi^{1/2} Y_i^* e_l\|_H^2
\end{aligned}$$

$$= \sum_{i=0}^{n-1} \mathbb{P}(A_i)\,((t_{i+1} \wedge t) - (t_i \wedge t))\; |Q_\xi^{1/2} Y_i^*|^2_{L_2(H)}$$

$$= \sum_{i=0}^{n-1} \mathbb{P}(A_i)\,((t_{i+1} \wedge t) - (t_i \wedge t))\; |Q_\xi^{1/2} Y_i|^2_{L_2(H)}$$

$$= \mathbb{E}\left[\int_0^t |Q_\xi^{1/2} Y(s)|^2_{L_2(H)}\, \mathrm{d}s\right]. \qquad \square$$

In the sequel we construct the stochastic integral on the Hilbert space H

Fix $T > 0$. It can be easily verified that on the space of simple processes $\mathscr{S}(H)$ the mapping

$$\|Y\|_T^2 := \mathbb{E}\left[\int_0^T |Q_\xi^{1/2} Y(s)|^2_{L_2(H)}\, \mathrm{d}s\right], \qquad Y \in \mathscr{S}(H),$$

is a seminorm on $\mathscr{S}(H)$. We define on $\mathscr{S}(H)$ the equivalence relation

$$Y \sim Z, \quad Y, Z \in \mathscr{S}(H) \quad :\Leftrightarrow \quad \|Y - Z\|_T = 0$$

and consider $\tilde{\mathscr{S}}(H) := \mathscr{S}(H)/\sim$. We now set

$$\mathscr{L}^2_{\xi,T}(H) := \overline{\tilde{\mathscr{S}}(H)}^{\|\cdot\|_T}.$$

We can extend the stochastic integral operator in (3.8) from simple processes to integrands in $\mathscr{L}^2_{\xi,T}(H)$.

Theorem 3.15. *Under the previous notation for any $T > 0$ and $t \in [0, T]$ there is a unique extension of I_t^ξ to a continuous operator*

$$I_t^\xi : (\mathscr{L}_{2,T}(H); \|\cdot\|_T) \to L^2(\Omega, \mathscr{F}, \mathbb{P}; H)$$

denoted by the same symbols.

For a proof see [PZ07], Theorem 8.7.

Definition 3.16. For $0 \leqslant t \leqslant T$ the operator I_t^ξ is called the *stochastic integral* with respect to ξ and for $Y \in \mathscr{L}^2_{\xi,T}(H)$ we introduce the notation

$$\int_0^t Y(s)\mathrm{d}\xi(s) := I_t^\xi(Y), \qquad t \geqslant 0.$$

Example 3.17. Let $A := \Delta = \frac{\partial^2}{\partial \zeta^2}$ be the second derivative in $H = H_0^1(0,1)$ and denote by $S := (S(t))_{t\geqslant 0}$ the $\mathscr{C}_0$-semigroup with generator A on H. Then $S\big|_{t\in[0,T]} \in \mathscr{L}_{2,T}(H)$ since for the orthonormal basis $(e_n)_{n\in\mathbb{N}}$ of eigenvectors in H with respect to A given by $e_n(\zeta) = \sin(\pi n \zeta)$ for $\zeta \in [0,1]$ and $n \in \mathbb{N}$ we have

$$
\begin{aligned}
\int_0^T \|Q_\xi^{1/2} S(s)\|_{L_2(H)}^2 \,\mathrm{d}s &= \int_0^T \sum_{n=1}^\infty \|Q_\xi^{1/2} e^{-(\pi n)^2 s} e_n\|_H^2 \,\mathrm{d}s \\
&= \sum_{n=1}^\infty \int_0^T e^{-2(\pi n)^2 s} \,\mathrm{d}s \, \|Q_\xi^{1/2} e_n\|_H^2 \\
&\leqslant \sum_{n=1}^\infty \frac{1}{2} \left(\frac{\|Q_\xi^{1/2} e_n\|_H}{\pi n} \right)^2 \left(1 - e^{-2(\pi n)^2 T}\right) \\
&\leqslant \frac{1}{\pi^2} |Q_\xi^{1/2}|_{L_2(H)}^2 < \infty.
\end{aligned}
$$

3.3 The Stochastic Convolution with Lévy Noise

In this section we collect the most important properties of the stochastic convolution, which plays a crucial role for the notion of mild solution of the Chafee–Infante equation in the sequel, and is a key tool in Chap. 4 of the main part of this work. Due to the previous section for each $T > 0$ the stochastic integral

$$
\int_0^T S(T-s)\,\mathrm{d}\xi(s)
$$

is well-defined. In the sequel we will show that the process $t \mapsto \int_0^t S(t-s)\,\mathrm{d}\xi(s)$ has a càdlàg version. We explain the notion of a variational generator, and demonstrate that it generates a semigroup of generalized contractions, for which a càdlàg version exists.

Definition 3.18. For a separable Hilbert space H a linear, unbounded, closed operator $A : D(A) \to H$ is called *variational* if

1. there exists a Hilbert space $V \hookrightarrow H$ densely imbedded, a continuous bilinear form $a : V \times V \to \mathbb{R}$ and constants $\alpha > 0$ and $c_0 \geqslant 0$ such that

$$
-a(v,v) \geqslant \alpha |v|_V^2 - c_0 \|v\|^2 \qquad \text{for all } v \in V, \tag{3.9}
$$

2. $D(A) = \{v \in V \mid a(v,\cdot)$ is continuous with respect to the topology in $H\}$,
3. $a(u,v) = \langle Au, v\rangle$ for all $u \in D(A)$ and $v \in V$.

For the definition of an analytic semigroup we refer to [DZ92], Appendix A.4, from which we cite the following proposition.

Proposition 3.19. *Let A be a variational generator in H such that inequality (3.9) is satisfied. Then A is the generator of an analytic semigroup $(S(t))_{t\geqslant 0}$ such that*

$$|S(t)|_{L(H)} \leqslant e^{ct}, \quad t \geqslant 0.$$

If A is symmetric, then A is self-adjoint.

The following proposition shows that the existence of a variational generator is sufficient for the existence of a càdlàg version.

Proposition 3.20. *Let ξ be a square integrable martingale in the separable Hilbert space H and $(S(t))_{t\geqslant 0}$ a semigroup of generalized contractions of H, which means there is an exponent $c \in \mathbb{R}$ such that*

$$\|S(t)\|_{L(H)} \leqslant e^{ct}, \qquad t \geqslant 0.$$

Then the process

$$\int_0^t S(t-s)\, \mathrm{d}\xi(s)$$

has a càdlàg version in H.

For the proof see [PZ07], p. 158. It is based on the so-called Kotelenez inequality for convolutions of evolution operators. We can finally verify that Δ is variational in H.

Example 3.21. We consider $H = H_0^1(0,1)$ with $\|h\|^2 = \int_0^1 (\nabla h)^2(\zeta)\, \mathrm{d}\zeta$ and

$$V = H_0^2(0,1) = \{v \in H_0^1(0,1) \mid \nabla v \in H_0^1(0,1)\}$$

with norm $|v|_V^2 = \int_0^1 (\Delta v)^2(\zeta)\, \mathrm{d}\zeta$ and $A = \Delta$. We show that A is a variational generator in H.

1. For

$$a(u,v) := -\int_0^1 (\Delta v)(\zeta)(\Delta u)(\zeta)\, \mathrm{d}\zeta, \quad u, v \in V,$$

we obtain for $v \in H_0^2(0,1)$

$$-a(v,v) = |v|_V^2,$$

such that in this case inequality (3.9) is fulfilled with $\alpha = 1$ and $c_0 = 0$.

2. The domain $D(A)$ can be identified with

$$D(A) = H^3(0,1) \cap H_0^2(0,1),$$

since for $u \in H^3(0,1) \cap H_0^2(0,1)$ and $v \in V$

$$\begin{aligned} a(u,v) &= -\int_0^1 (\Delta u)(\zeta)(\Delta v)(\zeta)\,\mathrm{d}\zeta = \int_0^1 (\nabla \Delta u)(\zeta)(\nabla v)(\zeta)\,\mathrm{d}\zeta \\ &\leqslant \left(\int_0^1 (\nabla \Delta u)^2(\zeta)\,\mathrm{d}\zeta\right) \int_0^1 (\nabla v)^2(\zeta)\,\mathrm{d}\zeta = |u|^2_{H^3 \cap H_0^2} \|v\|^2. \end{aligned}$$

3. The operator A is defined via the bilinear form by integration by parts through

$$a(u,v) = \int_0^1 (\nabla \Delta u)(\zeta)(\nabla v)(\zeta)\,\mathrm{d}\zeta, \qquad u \in D(A), v \in H.$$

Hence Δ is variational in H.

Example 3.22. In particular, for ξ^ε for $\varepsilon > 0$ and $\rho \in (0,1)$ defined in (3.6), the processes

$$\xi^*(t) = \int_0^t S(t-s) d\xi^\varepsilon(s), \qquad \varepsilon > 0, t \geqslant 0, \tag{3.10}$$

posses càdlàg versions.

3.4 The Stochastic Chafee–Infante Equation with Lévy Noise

In this section we will show existence and uniqueness for the global solution of the stochastic Chafee–Infante equation in the mild sense. For clarity in this section we will denote by $|\cdot|$ the modulus in $\mathbb{R}$, by $|\cdot|_{L^2}$ the norm in $L^2(0,1)$ and as before by $\|\cdot\|$ the norm in $H = H_0^1(0,1)$.

In preparation of the existence theorem we next define exit times of an H-valued process from large balls.

Definition 3.23. Let $X = (X(t))_{t \geqslant 0}$ be a an adapted càdlàg process in H. For $R > 0$ and $x \in B_R(0) \subset H$ we define the first exit time

$$\tau_R(x; X) := \inf\{t > 0 \mid \|X(t)\| > 2R\}.$$

Remark 3.24. The hitting times τ_R are stopping times (see for example [Kal97], Lemma 7.6.).

To prove existence and uniqueness for the stochastic Chafee–Infante equation, we shall proceed in two steps. First, we consider the deterministic system forced by the small noise component alone. In a second step, large jumps will be admitted.

We show that the mild solutions of the Chafee-Infante equation driven by small noise has a Càdlàg version in H

Let $\varepsilon > 0$. Consider the formal system (1.2) driven by $(\xi^{\varepsilon}(t))_{t\geqslant 0}$ instead of L

$$\begin{aligned} \frac{\partial}{\partial t}Y^{\varepsilon}(t,\zeta) &= \Delta Y^{\varepsilon}(t,\zeta) + f(Y^{\varepsilon}(t,\zeta)) + \varepsilon\dot{\xi}^{\varepsilon}(t,\zeta),\ \zeta \in [0,1],\ t > 0, \\ Y^{\varepsilon}(t,0) &= Y^{\varepsilon}(t,1) = 0, \qquad t > 0, \\ Y^{\varepsilon}(0,\zeta) &= x(\zeta), \qquad \zeta \in [0,1]. \end{aligned} \tag{3.11}$$

Let us first define the concept of mild solution in this context.

Definition 3.25. Denote by $(S(t))_{t\geqslant 0}$ the $\mathscr{C}^0$-semigroup generated by the second derivative $\Delta = \frac{\partial^2}{\partial \zeta^2}$ over $(0,1)$ with Dirichlet boundary conditions in H. Then for any time horizon $T > 0$ and $x \in H$ a (local) mild solution of (3.11) is a progressively measurable process $(Y^{\varepsilon}(t))_{t\in[0,T]}$ in H fulfilling for all $t \in [0,T]$ the integral equation

$$Y^{\varepsilon}(t) = S(t)x + \int_0^t S(t-s)f(Y^{\varepsilon}(s))\,\mathrm{d}s + \varepsilon\int_0^t S(t-s)\mathrm{d}\xi^{\varepsilon}(s) \qquad \mathrm{d}\zeta \otimes \mathbb{P}\text{-a.s.} \tag{3.12}$$

A global mild solution is a mild solution existing for all $t \geqslant 0$.

We next recall the existence of a local mild solution based on results in [PZ07].

Proposition 3.26 (Existence of a unique local mild solution). *For any $R > 0$, $x \in B_R(0)$, $T > 0$, and $\varepsilon > 0$ there is a unique local mild solution of (3.11), which satisfies for any $t \in [0, T \wedge \tau_R(x, Y^{\varepsilon})]$*

$$Y^{\varepsilon}(t) = S(t)x + \int_0^t S(t-s)f(Y^{\varepsilon}(s))\,\mathrm{d}s + \varepsilon\int_0^t S(t-s)\mathrm{d}\xi^{\varepsilon}(s). \tag{3.13}$$

Moreover Y^{ε} has a càdlàg version.

Proof. First note that since $(L(t))_{t\geqslant 0}$ is a symmetric pure jump Lévy process (see Definition 3.9) for any $\rho \in (0,1)$ and $\varepsilon > 0$ the process ξ^{ε} defined in (3.6), Sect. 3.1, is a mean zero martingale with moments of second order. We denote by $Q_{\xi^{\varepsilon}}$ the trace class operator given by Theorem 3.12. We verify the assumptions of Theorem 9.29 in [PZ07]. By Lemma 2.1 and (2.5), for all $u, v \in B_R(0)$ and $t \geqslant 0$ it follows

$$\|S(t)(f(u) - f(v))\| \leqslant \|S(t)\|\|f(u) - f(v)\| \leqslant e^{-c_0 t} K_R \|u - v\|$$

and in particular

$$\|S(t)f(u)\| \leqslant e^{-c_0 t} K_{1,R}\|u\|, \qquad t \geqslant 0, u \in B_R(0),$$

where $K_{1,R}$ is defined by (2.3). Take an orthonormal basis $(e_n)_{n\in\mathbb{N}}$, then the set $(Q_{\xi^\varepsilon}^{1/2}(e_n)_{n\in\mathbb{N}})$ is an orthonormal basis in $Q_{\xi^\varepsilon}^{1/2}(H)$. We may calculate for $t \geqslant 0$

$$\|S(t)\|^2_{L_2(Q_{\xi^\varepsilon}^{1/2}(H);H)} = \sum_{k=1}^{\infty} \|S(t)Q_{\xi^\varepsilon}^{1/2}e_k\|^2 \leqslant e^{-c_0 t} \sum_{k=1}^{\infty} \|Q_{\xi^\varepsilon}^{1/2}e_k\|^2 \leqslant \operatorname{Tr} Q_{\xi^\varepsilon}$$

Thus

$$\int_0^T \|S(t)\|^2_{L_2(Q_{\xi}^{1/2}(H);H)} \, \mathrm{d}t \leqslant T \operatorname{Tr} Q_\xi$$

Hence we may apply Theorems 9.29 and 9.15 in Chap. 9 of [PZ07], which guarantee under these assumptions the existence of a local unique weak solution, resp. the existence of an equivalent local mild solution of (3.11). If $\xi^*(t) = \int_0^t S(t-s)\xi^\varepsilon(s)$ and $Y^\varepsilon(t) = v(t) + \varepsilon\xi^*(t)$, we can rewrite equation (3.11) as

$$v(t) = S(t)x + \int_0^t S(t-s) f(v(s) + \varepsilon\xi^*(s-)) \, \mathrm{d}s, \quad t \geqslant 0.$$

We apply Proposition 3.20 to Example 3.21, which implies that $\psi := \varepsilon\xi^*$ has a càdlàg version. Clearly v is continuous in H, therefore $t \to Y^\varepsilon(t) = v(t) + \varepsilon\xi^*(t)$ inherits the càdlàg property on the stochastic interval $[0, T \wedge \tau_R(x, Y^\varepsilon)]$. □

Proposition 3.27. *There exists a global unique mild solution $(Y^\varepsilon(t))_{t\geqslant 0}$ of (3.11), which has a càdlàg version.*

Proof. Below, we prove that $\sup_{[0,\tau_R(x;Y^\varepsilon)]} \|Y^\varepsilon(t)\|$ is bounded a.s. independently on R. This implies that $\tau_R(x, Y^\varepsilon) \to \infty$ a.s. when $R \to \infty$ so that the solution constructed in Proposition 3.26 is global.

We use $v(t) = Y^\varepsilon(t) - \varepsilon\xi^*(t)$ which satisfies

$$\frac{\mathrm{d}}{\mathrm{d}t} v(t) = \Delta v(t) + f(v(t) + \varepsilon\xi^*(t)).$$

We multiply by $-\Delta v$ and integrate to obtain after simple computations:

$$\frac{1}{2}\frac{\mathrm{d}}{\mathrm{d}t}\|v(t)\|^2 + |\Delta v(t)|^2 + 3\lambda \int_0^1 (v(t))^2 (\nabla v(t))^2 \, d\zeta$$

$$= -\lambda \int_0^1 6\varepsilon\xi^*(t)v(t) (\nabla v(t))^2 + 3\varepsilon\nabla\xi^*(t)v(t)^2\nabla v(t) + 3\varepsilon^2\xi^*(t)^2 (\nabla v(t))^2$$

$$+ 6\varepsilon^2\xi^*(t)\nabla\xi^*(t)v(t)\nabla v(t) + 3\varepsilon^3\xi^*(t)^2\nabla\xi^*(t)\nabla v(t)$$
$$- (\nabla v(t))^2 - \varepsilon\nabla\xi^*(t)\nabla v(t)\, d\zeta.$$

In fact the above equality should be justified by an adequate regularisation argument.

We control each term of the right hand side. For the first one, we write by Cauchy–Schwarz inequality:

$$-\lambda\int_0^1 6\varepsilon\xi^*(t)v(t)\left(\nabla v(t)\right)^2\, d\zeta$$
$$\leq 6\varepsilon\lambda\left(\int_0^1 (v(t))^2\left(\nabla v(t)\right)^2 d\zeta\right)^{1/2}\left(\int_0^1 \xi^*(t)^2\left(\nabla v(t)\right)^2 d\zeta\right)^{1/2}$$

and deduce thanks to the domination of the L^∞ norm by the H norm:

$$-\lambda\int_0^1 6\varepsilon\xi^*(t)v(t)\left(\nabla v(t)\right)^2\, d\zeta \leq \frac{\lambda}{2}\int_0^1 (v(t))^2\left(\nabla v(t)\right)^2 d\zeta + C\,\|\xi^*(t)\|^2\|v(t)\|^2.$$

The three next admit the same bound with a possibly different constant. Then, again by Cauchy–Schwarz inequality,

$$-3\lambda\varepsilon^3\int_0^1 \xi^*(t)^2\nabla\xi^*(t)\nabla v(t)d\zeta \leqslant 3\lambda\varepsilon^3|\xi^*(t)|_\infty^2\|\xi^*(t)\|\;\|v(t\|$$
$$\leqslant \|v(t)\|^2 + \frac{3}{4}\lambda^2\varepsilon^6\|\xi^*(t)\|^6$$

The last two term are easy to treat and, neglecting the positive terms on the left hand side, we obtain a constant $\tilde{C}$ depending on λ and ε such that:

$$\frac{1}{2}\frac{\mathrm{d}}{\mathrm{d}t}\|v(t)\|^2 \leqslant \tilde{C}\left(\|\xi^*(t)\|^2 + 1\right)\left(\|v(t)\|^2 + 1\right).$$

Therefore, by Gronwall Lemma,

$$\|v(t)\|^2 \leqslant e^{\int_0^t\left(\|\xi^*(s)\|^2+1\right)ds}\left(\|x\|^2 + 1\right).$$

Since ξ^* is a càdlàg process with values in H, we have for any T

$$\sup_{t\in[0,T]}\|\xi^*(t)\| < \infty,\ a.s.$$

We deduce that $\sup_{[0,\tau_R(x;Y^\varepsilon)]} \|Y^\varepsilon(t)\|$ is bounded a.s. independently on R as claimed. The càdlàg property is inherited from the sequence of the unique local mild solutions. □

Combining the previous results we prove the global existence, uniqueness and the existence of a strong solution of the stochastic Chafee-Infante equation (1.2) driven by infinite-dimensional α-stable noise.

In a second step, we shall now add the big jumps in a controlled way to the solutions of the Chafee–Infante equation with small noise. For $\varepsilon > 0$ recall from Sect. 3.1 the large jump component η^ε which is a compound Poisson process with intensity

$$\beta_\varepsilon := \nu\left(\frac{1}{\varepsilon^\rho} B_1^c(0)\right),$$

jump probability measure ν outside the ball $\frac{1}{\varepsilon^\rho} B_1(0)$ given by (3.5), and jump times defined recursively by

$$T_0 := 0, \qquad T_k := \inf\left\{t > T_{k-1} \;\middle|\; \|\Delta_t L\| > \varepsilon^{-\rho}\right\}, \quad k \geqslant 1,$$

with interjump times

$$t_0 = 0, \qquad t_k := T_k - T_{k-1}, \quad k \geqslant 1,$$

with exponential laws $\mathscr{L}(t_k) = EXP(\beta_\varepsilon)$, and big jumps W_k at time T_k, $k \geqslant 1$.

Definition 3.28. The mild solution for the system (1.2) with time horizon $T > 0$ is the progressively measurable process $(X^\varepsilon(t))_{t\in[0,T]}$ fulfilling for $t \in [0,T]$

$$X^\varepsilon(t) = S(t)x + \int_0^t S(t-s) f(X^\varepsilon(s))\,\mathrm{d}s + \int_0^t S(t-s)\mathrm{d}\xi^\varepsilon(s) + \int_0^t S(t-s)\mathrm{d}\eta^\varepsilon(s).$$

Its global solution is a solution defined for all times $t \geqslant 0$.

Since, as was remarked earlier, the big jumps of the mild solution are just the big jumps of η^ε, this equation is easily seen to have a unique global solution. It is recursively defined between big jump times of η^ε by the increments of the solution of the Chafee–Infante equation perturbed by the small jump component studied before. More formally, for $k \geqslant 1$ and $T_k \leqslant t < T_{k+1}$ we have

$$X_t^\varepsilon - X_{T_k}^\varepsilon = Y_{t-T_k}^\varepsilon, \quad \Delta X_{T_k}^\varepsilon = W_k,$$

where Y^ε is the global solution of the Chafee–Infante equation with small noise perturbation.

3.5 The Strong Markov Property

In this section we sketch a proof of the strong Markov property of our solutions Y^ε of (3.11), a crucial ingredient for our method to describe exit times. We mostly proceed along the lines of reasoning established in [DZ92] and [PZ07]. But since our setting is not covered explicitly there, we prefer to indicate the arguments needed.

Definition 3.29. Let $(\mathscr{F}_t)_{t\geqslant 0}$ be a complete right-continuous filtration. An $(\mathscr{F}_t)$-adapted process $(X(t))_{t\geqslant 0}$ with values in a measurable space $(E, \mathscr{E})$ has the *Markov property* if it satisfies for $0 \leqslant s \leqslant t$

$$\mathbb{P}(X(t) \in A \mid \mathscr{F}_s) = \mathbb{P}(X(t) \in A \mid X(s)) \qquad \text{for all } A \in \mathscr{E}. \tag{3.14}$$

For a separable Hilbert space $E = H$ and the Borel-σ-Algebra $\mathscr{E} = \mathscr{B}(H)$ we denote $B_b(H)$ the space of real-valued, bounded Borel functions equipped with the norm

$$|f|_{B_b} = \sup_{x\in H} |f(x)|_H.$$

We denote $Y^\varepsilon(t; s, x)$ the value of the global mild solution of (3.11) obtained in Proposition 3.27 at time $t \geqslant 0$ starting at time $0 \leqslant s \leqslant t$ in $x \in H$. For $\varphi \in B_b(H)$, $0 \leqslant s \leqslant t$ and $x \in H$ define by

$$(P_{s,t}\varphi)(x) := \mathbb{E}\left[\varphi(Y^\varepsilon(t; s, x))\right]$$

the *transition operator of* $(Y^\varepsilon(t))_{t\geqslant 0}$. Note that in this case for $A \in \mathscr{B}(H)$

$$\mathbb{E}\left[\mathbf{1}_A(Y^\varepsilon(t; s, x))\right] = \mathbb{P}(Y^\varepsilon(t; s, x) \in A).$$

Recall that the family $(P_{s,t})$ is called *homogeneous*, if the transition operators are translation invariant, i.e. $P_{s,t}\varphi = P_{0,t-s}$ for all $0 \leqslant s \leqslant t$. Also recall that the family $(P_{s,t})$ satisfies the *Feller property*, if for $\varphi \in \mathscr{C}_b(H)$ and $0 \leqslant s \leqslant t \leqslant T$ the map

$$P_{s,t}\varphi : H \to \mathbb{R}, \qquad x \mapsto (P_{s,t}\varphi)(x)$$

is continuous. We shall start with the observation that our global mild solutions of the Chafee–Infante equation Y^ε, $\varepsilon > 0$, possess families of homogeneous transition operators endowed with the Feller property. In this case we say that the process $(Y^\varepsilon(t))_{t\geqslant 0}$ is homogeneous and has the Feller property.

Proposition 3.30. *The mild solution Y^ε of (3.11) is a homogeneous Markov process possessing the Feller property.*

Proof. We refer to [PZ07], Theorems 9.29 and 9.30 and Remark 9.33. There the authors prove that mild solutions of SPDEs with time independent Lipschitz coefficients driven by an additive mean zero Lévy martingale with second moments in H are homogeneous Markov processes endowed with the Feller property. This shows that the process Y_R^ε, solution of (3.11) with f replaced by a truncated nonlinear term equal to f inside the ball $B_R(0)$ and equal to 0 outside $B_{2R}(0)$, is a homogeneous Markov process possessing the Feller property. It is to prove that Y^ε is also a homogeneous Markov process. We then observe that

$$\begin{aligned}|\mathbb{E}[\varphi(Y_R^\varepsilon(t))] - \mathbb{E}[\varphi(Y^\varepsilon(t))]| &= \mathbb{E}[|\varphi(Y_R^\varepsilon(t)) - \varphi(Y^\varepsilon(t))|\mathbf{1}_{\{t>\tau_R(x;Y^\varepsilon)\}}] \\ &\leqslant 2\|\varphi\|_\infty \mathbb{P}(t > \tau_R(x;Y^\varepsilon))\end{aligned}$$

so that, using the estimate in the proof of Proposition 3.27, we deduce that the transition operator of Y_R^ε converges to the transition operator associated to Y^ε uniformly on any balls in H. We deduce that Y^ε also possesses the Feller property. □

Lemma 3.31. *For all $\varphi \in B_b(H)$, $0 \leqslant r \leqslant s \leqslant t$, and $\mathscr{F}_s$ measurable random variables $\tilde{X} \in L^2(\Omega, \mathscr{F}_s, \mathbb{P}; H)$ we have*

$$\mathbb{E}\left[\varphi(Y^\varepsilon(t;r,\tilde{X})) \mid \mathscr{F}_s\right] = (P_{s,t}\varphi)(Y^\varepsilon(s;r,\tilde{X})) \qquad \mathbb{P}\text{-a.s.} \tag{3.15}$$

Proof. Each bounded measurable function φ can be approximated monotonically by a sequence of simple functions $(\varphi_n)_{n\in\mathbb{N}}$. This allows to pass to the limit in (3.14). □

All properties discussed so far only concern the finite dimensional marginals of Y^ε. In order to prove the strong Markov property one has to pass to laws on path space. We follow here [DZ92], Chap. 9.2.

Proposition 3.32. *Let $\varphi_1, \ldots, \varphi_n \in B_b(H)$, $0 \leqslant s \leqslant t$ and $0 \leqslant h_1 \leqslant \cdots \leqslant h_n$ be arbitrary. Then for the solution $(Y^\varepsilon(t;s,\tilde{X}))_{t\geqslant s}$ of (3.11) with initial values $(s,\tilde{X})$ the relation*

$$\begin{aligned}&\mathbb{E}\left[\varphi_1(Y^\varepsilon(t+h_1;s,\tilde{X}))\,\varphi_2(Y^\varepsilon(t+h_2;s,\tilde{X}))\,\ldots\,\varphi_n(Y^\varepsilon(t+h_n;s,\tilde{X})) \mid \mathscr{F}_t\right] \\ &= Q_{h_1,\ldots,h_n}^{\varphi_1,\ldots,\varphi_n}(t, Y^\varepsilon(t;s,\tilde{X})) \qquad \mathbb{P}\text{-a.s.}\end{aligned} \tag{3.16}$$

is valid, where $Q_{h_1,\ldots,h_n}^{\varphi_1,\ldots,\varphi_n} : \mathbb{R}_+ \times H \to \mathbb{R}$ is recursively defined by

$$Q_{h_1,\ldots,h_n}^{\varphi_1,\ldots,\varphi_n}(s,x) := P_{s,s+h_1}(\varphi_1 \cdot Q_{h_2,\cdots,h_n}^{\varphi_2,\cdots,\varphi_n}(s+h_1;\cdot))(x), \quad Q_{h_1}^{\varphi_1}(s,x) = P_{s,s+h_1}\varphi_1(x). \tag{3.17}$$

$Q_{h_1,\ldots,h_n}^{\varphi_1,\ldots,\varphi_n}$ is a Borel function.

This can be proved by induction over n identically to Proposition 9.11 in [DZ92], p. 252.

Since Y^ε has almost surely càdlàg trajectories, the theory differs slightly from the case of continuous trajectories treated in [DZ92], Chap. 9.2.

Definition 3.33. Denote by $D := D([0,\infty); H)$ the space of càdlàg functions with values in H, and by $\mathbf{P}^{s,x}$ the distribution of $Y^\varepsilon(s + \cdot\ ; s, x)$ on $(D, \mathscr{B}(D))$ with respect to the Skorohod topology. It is defined by

$$\mathbf{P}^{s,x}(\mathscr{A}) = \mathbb{P}(Y^\varepsilon(s + \cdot\ ; s, x) \in \mathscr{A}) \qquad \text{for } \mathscr{A} \in \mathscr{B}(D).$$

A *cylindrical set* $\mathscr{Z}$ in D is defined by $0 \leqslant h_1 \leqslant \cdots \leqslant h_n$ and $A_1, \ldots, A_n \in \mathscr{B}(H)$

$$\mathscr{Z} = \mathscr{Z}(h_1, \ldots, h_n; A_1, \ldots, A_n) = \{g \in D \mid g(h_1) \in A_1, \ldots, g(h_n) \in A_n\}.$$

Remark 3.34. 1. The measure $\mathbf{P}^{s,x}$ is uniquely determined by its values on cylindrical sets.
2. For a cylindrical set in D over H of the above form it follows by definition

$$\mathbf{P}^{s,x}(\mathscr{Z}) = \mathbb{P}\big(Y^\varepsilon(s + h_1; s, x) \in A_1, \ldots, Y^\varepsilon(s + h_n; s, x) \in A_n\big).$$

In particular by the Chapman–Kolmogorov equation we have

$$\mathbf{P}^{s,x}(\mathscr{Z}) = Q_{h_1,\ldots,h_n}^{\mathbf{1}_{A_1},\ldots,\mathbf{1}_{A_n}}(s, x).$$

Hence (3.16) has for $0 \leqslant s \leqslant t$ the shape

$$\mathbb{P}(Y^\varepsilon(t + \cdot\ ; s, \tilde{X}) \in \mathscr{Z} \mid \mathscr{F}_t) = \mathbf{P}^{t, Y^\varepsilon(t; s, \tilde{X})}(\mathscr{Z}). \tag{3.18}$$

Definition 3.35. We denote by $\mathbf{E}^{s,x} := \mathbb{E}_{\mathbf{P}^{s,x}}, s \geqslant 0, x \in H$.

Remark 3.36. With this notation (3.18) can be rewritten as

$$\mathbb{E}\left[\mathbf{1}\{Y^\varepsilon(t + \cdot \quad s, \tilde{X}) \in \mathscr{Z}\} \mid \mathscr{F}_s\right] = \mathbf{E}^{t, Y^\varepsilon(t; s, \tilde{X})}[\mathbf{1}_{\mathscr{Z}}]. \tag{3.19}$$

Since each measurable, bounded function $\Psi : D \to \mathbb{R}$ can be approximated monotonically by simple functions, the identity

$$\mathbb{E}\left[\Psi(Y^\varepsilon(t + \cdot\ ; s, \tilde{X})) \mid \mathscr{F}_s\right] = \mathbf{E}^{t, Y^\varepsilon(t; s, \tilde{X})}[\Psi]. \tag{3.20}$$

follows.

Definition 3.37. Let τ be a $(\mathscr{F}_t)_{t \geqslant 0}$-stopping time. We denote by

$$\mathscr{F}_\tau := \sigma\{A \in \mathscr{F} \mid \{\tau \leqslant t\} \cap A \in \mathscr{F}_t\}$$

the σ-algebra of the τ-*past*. We say that Y^ε satisfies the strong Markov property if for each $s \geqslant 0$, stopping time $\tau \geqslant s$, $\tilde{X}$ a $\mathscr{F}_s$-measurable random variable and measurable mapping $\Psi : (D, \mathscr{B}(D)) \to \mathbb{R}$ we have

$$\mathbb{E}\left[\Psi(Y^\varepsilon(\tau + \cdot\ ; s, \tilde{X})) \mid \mathscr{F}_\tau\right] = \mathbf{E}^{s, Y^\varepsilon(\tau; s, \tilde{X})}[\Psi] \qquad \mathbb{P}(\cdot \mid \tau < \infty)\text{-a.s.}$$

Proposition 3.38. *The mild solution Y^ε of (3.11) satisfies the strong Markov property.*

Proof. We have to show that for all nonnegative Borel functions $\Psi : (D, \mathscr{B}(D)) \to \mathbb{R}$, $s \geq 0$, $\tilde{X} \in L^2(\Omega, \mathscr{F}_s, \mathbb{P}; H)$ and all stopping times $\tau \geq s$ and $A \in \mathscr{F}_\tau$ the equation

$$\mathbb{E}\left[\Psi(Y^\varepsilon(\tau + \cdot\,; s, \tilde{X}))\mathbf{1}_{A\cap\{\tau<\infty\}}\right] = \mathbb{E}\left[\mathbf{E}^{\tau, Y^\varepsilon(\tau; s, \tilde{X})}\left[\Psi\, \mathbf{1}_{A\cap\{\tau<\infty\}}\right]\right]$$

holds. Each stopping time τ has for fixed $2 \leq q \in \mathbb{N}$ and $N \in \mathbb{N}$ a q-adic approximation $\tau_N := \frac{\lfloor \tau q^N \rfloor + 1}{q^N}$ with countably many values. Since $\tau_N \geq \tau$ we have $\mathscr{F}_{\tau_N} \supset \mathscr{F}_\tau$ and $\tau^N \searrow \tau$ for $N \to \infty$ $\mathbb{P}$-a.s. For Markov processes in Polish spaces such as H with a countable set of time parameters I, which is closed under summation, the Markov property implies the strong Markov property, see for example [Kle05], p. 352. In our case this property is fulfilled for $(Y^\varepsilon(t))_{t\in I_n}$, where $I_n = \{\frac{k}{q^n}, k \in \mathbb{N}\}$. Hence we have for $n \in \mathbb{N}$

$$\mathbb{E}\left[\Psi(Y^\varepsilon(\tau_n + \cdot\,; s, \tilde{X}))\mathbf{1}_{A\cap\{\tau<\infty\}}\right] = \mathbb{E}\left[\mathbf{E}^{\tau_n, Y^\varepsilon(\tau_n; s, \tilde{X})}\left[\Psi\, \mathbf{1}_{A\cap\{\tau<\infty\}}\right]\right]. \tag{3.21}$$

It remains to pass to the limit $n \to \infty$ in (3.21). We denote by

$$\mathscr{M} := \{\Psi : (D, \mathscr{B}(D)) \to \mathbb{R} \mid \Psi \text{ measurable }\}.$$

Clearly for each $h \geq 0$ and $\varphi \in \mathscr{C}_b(H)$ the point evaluations

$$\Psi(f) := \varphi(f(h))$$

are a subclass of $\mathscr{M}$. For point evaluations of this type equation (3.21) has the form

$$\mathbb{E}\left[\varphi(Y^\varepsilon(\tau_n + h; s, \tilde{X}))\mathbf{1}_{A\cap\{\tau<\infty\}}\right] = \mathbb{E}\left[(P_{\tau_n, \tau_n + h}\varphi)(Y^\varepsilon(\tau_n; s, \tilde{X}))\mathbf{1}_{A\cap\{\tau<\infty\}}\right]. \tag{3.22}$$

Since Y^ε possesses the Feller property, for each $\varphi \in \mathscr{C}_b(H)$ and $t, h > 0$ the mapping

$$H \to \mathbb{R}, \qquad x \mapsto (P_{t,t+h}\varphi)(x)$$

is continuous. It can be shown analogously to [DZ92], Theorem 9.1, that for any $t \geq 0$ and $\tilde{X} \in L^2(\Omega, \mathscr{F}_t, \mathbb{P}; H)$ the mapping $h \mapsto \mathbb{E}[(Y^\varepsilon(t + h; t, \tilde{X}))^2]$ is continuous and hence that for each $\varphi \in \mathscr{C}_b(H)$, $x \in H$ and $t > 0$ the mapping

$$h \mapsto (P_{t,t+h}\varphi)(x)$$

is continuous. In addition, by Theorem 3.27, Y^ε has almost surely right-continuous trajectories. Hence we can pass to the limit and obtain

$$\mathbb{E}\left[\varphi(Y^{\varepsilon}(\tau+h;s,\tilde{X}))\mathbf{1}_{A\cap\{\tau<\infty\}}\right] = \mathbb{E}\left[(P_{\tau,\tau+h}\varphi)(Y^{\varepsilon}(\tau;s,\tilde{X}))\mathbf{1}_{A\cap\{\tau<\infty\}}\right]. \tag{3.23}$$

This is equivalent to

$$\mathbb{E}\left[\varphi(Y^{\varepsilon}(\tau+h;s,\tilde{X})) \mid \mathscr{F}_{\tau}\right] = (P_{\tau,\tau+h}\varphi)(Y^{\varepsilon}(\tau;s,\tilde{X})) \quad \mathbb{P}(\cdot \mid \tau<\infty)\text{-a.s.} \tag{3.24}$$

Since $\mathbf{1}_B$ for $B \in \mathscr{B}(D)$ can be approximated by continuous functions, and measurable functions can be approximated by simple functions, we obtain even for $\varphi \in B_b(H)$ that

$$\mathbb{E}\left[\varphi(Y^{\varepsilon}(\tau+h;s,\tilde{X})) \mid \mathscr{F}_{\tau}\right] = (P_{\tau,\tau+h}\varphi)(Y^{\varepsilon}(\tau;s,\tilde{X})) \quad \mathbb{P}(\cdot \mid \tau<\infty)\text{-a.s.} \tag{3.25}$$

By induction on $n \in \mathbb{N}$ we can now show the following analogue of Proposition 3.32 in which deterministic times are replaced with stopping times. Consider the solution $(Y^{\varepsilon}(t;s,\tilde{X}))_{t\geqslant s}$ of (3.11) with initial values $(s,\tilde{X})$. For $\varphi_1,\ldots,\varphi_n \in B_b(H)$, $s \geqslant 0$, a stopping time $\tau \geqslant s$ and $0 \leqslant h_1 \leqslant h_2 \leqslant \cdots \leqslant h_n$ we have

$$\begin{aligned}&\mathbb{E}\left[\varphi_1(Y^{\varepsilon}(\tau+h_1;s,\tilde{X}))\,\varphi_2(Y^{\varepsilon}(\tau+h_2;s,\tilde{X}))\,\ldots\,\varphi_n(Y^{\varepsilon}(\tau+h_n;s,\tilde{X})) \mid \mathscr{F}_{\tau}\right]\\ &= Q^{\varphi_1,\ldots,\varphi_n}_{h_1,\ldots,h_n}(\tau, Y^{\varepsilon}(\tau;s,\tilde{X})) \qquad \mathbb{P}(\cdot \mid \tau<\infty)\text{-a.s.,}\end{aligned} \tag{3.26}$$

where the right-hand side is defined by (3.16). Hence for a cylindrical set $\mathscr{Z}$ we can rewrite the last equation as

$$\mathbb{P}(Y^{\varepsilon}(\tau+\cdot\,;s,\tilde{X}) \in \mathscr{Z} \mid \mathscr{F}_{\tau}) = \mathbf{P}^{\tau,Y^{\varepsilon}(\tau;s,\tilde{X})}(\mathscr{Z}). \tag{3.27}$$

By monotone approximation of measurable and bounded $\Psi : D \to \mathbb{R}$ and with the convention $\mathbf{E}^{s,x} = \mathbb{E}_{\mathbf{P}^{s,x}}$ we obtain the desired equation

$$\mathbb{E}\left[\Psi(Y^{\varepsilon}(\tau+\cdot\,;s,\tilde{X})) \mid \mathscr{F}_{\tau}\right] = \mathbf{E}^{\tau,Y^{\varepsilon}(\tau;s,\tilde{X})}(\Psi).$$

This completes the proof. □

3.6 Basics on Slowly and Regularly Varying Functions

In this section we discuss basic properties of slowly varying functions that are needed in the definition of regularly varying Lévy measures. Regular variation plays a key role in the concept of Lévy measures for which the corresponding solutions of the stochastic Chafee–Infante equation have exit times which grow polynomially in the inverse noise amplitude ε in the small noise limit.

We cite from [HL06] and [BGT87].

Definition 3.39. A positive measurable function $\ell : (0,\infty) \to (0,\infty)$ satisfying

$$\lim_{x\to\infty} \frac{\ell(yx)}{\ell(x)} = 1 \qquad \text{for all } \ y > 0 \tag{3.28}$$

is called *slowly varying*.

Definition 3.40. A regularly varying function with index $-\beta$ for $\beta > 0$, is a nonnegative, measurable function $v : (0,\infty) \to (0,\infty)$ which satisfies for any $y > 0$

$$\lim_{x\to\infty} \frac{v(xy)}{v(x)} = y^{-\beta}.$$

The link between regular varying functions and slowly varying functions is discussed by ([BGT87], Theorem 1.4.1).

Theorem 3.41. *For any regularly varying function $v : (0,\infty) \to (0,\infty)$ with index β there is a slowly varying function $\ell : (0,\infty) \to (0,\infty)$ such that*

$$v(x) = x^{\beta}\ell(x) \quad \textit{for all } x \in (0,\infty).$$

Example 3.42. 1. $\ell_1(x) = \ln(x), x > 0$, defines a slowly varying function.
2. $\ell_2(x) = \exp(\ln(x)^{(1/3)}\cos(\ln(x)^{(1/3)})), x > 0$, defines a slowly varying function with infinite oscillations, i.e.

$$\liminf_{x\to\infty} \ell_2(x) = 0 \quad \text{and} \quad \limsup_{x\to\infty} \ell_2(x) = \infty.$$

These oscillations are, however, of vanishing order as we can see in the following proposition.

The following proposition is very important for our purposes. It tells us that in comparison to polynomials slowly varying functions have the same asymptotic behavior as constants.

Proposition 3.43. *A slowly varying function $\ell : (0,\infty) \to (0,\infty)$ has the property that for any $a > 0$*

$$\lim_{x\to\infty} x^{a}\ell(x) = \infty,$$

$$\lim_{x\to\infty} x^{-a}\ell(x) = 0.$$

For the proof see [BGT87], Theorem 1.2.1.

We next summarize basic results about regularly varying measures

So far we did not require any property from ν besides symmetry. From now on we shall concentrate on ν for which the tail decays asymptotically in the order of $r^{-\alpha}, \alpha \in (0,2)$, for $r \to \infty$. This is a natural generalization of the jump measure of α-stable processes with values in H. In order to describe the asymptotic polynomial increase in ε of the large jumps we introduce the following notions (see [BGT87], [HL06] and Appendix 3.6).

Let us extend the notion of regular variation to measures on a Hilbert space H.

Definition 3.44. 1. Let H be a separable Hilbert space. Denote by $M_0(H)$ the class of all Radon measures $\nu : \mathscr{B}(H) \to [0,\infty)$ with the property

$$\mu(A) < \infty \quad \Leftrightarrow \quad A \in \mathscr{B}(H),\ 0 \notin \bar{A},$$

where $\bar{A}$ stands for the closure of a set A in H.

2. A measure $\nu \in M_0(H)$ is *called regularly varying with index* $-\beta$ if there exists a non-zero measure $\mu \in M_0(H)$ and a regularly varying function v of index $-\beta$ such that

$$\lim_{t\to\infty} v(t)\nu(tA) = \mu(A) \qquad \text{for } A \in \mathscr{B}(H), 0 \notin \bar{A}.$$

We fix in all this work, and especially in Chaps. 5–7

- A symmetric, regularly varying measure $\nu \in M_0(H)$ with index $\alpha = -\beta \in (0,2)$ and limiting measure $\mu \in M_0(H)$.

By Theorem 3.41 this means that there is a slowly varying function $\ell : (0,\infty) \to (0,\infty)$ such that for all $A \in \mathscr{B}(H)$

$$\nu\,(tA)\,[t^{-\alpha}\ell(t)\mu(A)]^{-1} \to 1 \quad \text{as } t \to \infty.$$

In particular for $\varepsilon > 0$

$$\beta_\varepsilon = \nu\left(\frac{1}{\varepsilon^\rho} B_1^c(0)\right)$$

and

$$\beta_\varepsilon[\varepsilon^{\alpha\rho}\ell\left(\frac{1}{\varepsilon^\rho}\right)\mu(B_1^c(0))]^{-1} \to 1 \quad \text{as } \varepsilon \to 0+. \tag{3.29}$$

Example 3.45. The Lévy measure ν for an α-stable process in a Hilbert space H is introduced via spherical coordinates and has the shape

$$\nu(\mathrm{d}x) = \sigma(\mathrm{d}s)\frac{\mathrm{d}r}{r^{1+\alpha}},$$

where $s = x/\|x\|$ and $r = \|x\|$, and σ is an arbitrary finite measure on the unit sphere (see e.g. [AG79]). It is regularly varying with index $-\alpha$ with $\alpha \in (0, 2)$. For $t > 0$ and $A \in \mathscr{B}(H)$ with $0 \notin \bar{A}$ we may calculate by substitution for $t > 0$

$$\nu(tA) = \int_{tA} \nu(\mathrm{d}y) = \int_{tA} \sigma(s)\frac{\mathrm{d}r}{r^{1+\alpha}} = \int_{A} \sigma(s')\frac{t\mathrm{d}r'}{(tr')^{1+\alpha}} = \nu(A)t^{-\alpha}.$$

Hence in this case $\mu = \nu$ and $\ell = 1$.

In the following definition we collect for further reference the ε-dependent quantities that will turn out to be key for the description of the asymptotic behavior of the exit times and the metastable behavior of the solution processes x^{ε} of our stochastic Chafee–Infante equation.

Definition 3.46. We define the asymptotic weight of the tail the jump measure attributes to the reshifted domains of attraction (see Sects. 2.1.2 and 2.2.1) by

$$\lambda^{\pm}(\varepsilon) := \nu\left(\frac{1}{\varepsilon}\left(D_0^{\pm}\right)^c\right), \qquad \varepsilon > 0,$$

and the critical time scale for metastable behavior by

$$\lambda^{0}(\varepsilon) := \nu\left(\frac{1}{\varepsilon}B_1^c(0)\right), \qquad \varepsilon > 0.$$

Recall the closely related scale of the intensity of the large jumps

$$\beta_{\varepsilon} = \nu\left(\frac{1}{\varepsilon^{\rho}}B_1^c(0)\right), \qquad \varepsilon > 0.$$

The scales thus defined increase polynomially in ε in the limit $\varepsilon \to 0+$, since ν is regularly varying. More precisely, according to Theorem 3.41 there is a slowly varying function ℓ such that for any $\varepsilon > 0$ we have

$$\lambda^{\pm}(\varepsilon)\left[\varepsilon^{\alpha}\,\ell(1/\varepsilon)\,\mu\left((D_0^{\pm})^c\right)\right]^{-1} \to 1, \qquad \text{as } \varepsilon \to 0+,$$

$$\lambda^{0}(\varepsilon)\left[\varepsilon^{\alpha}\,\ell(1/\varepsilon)\,\mu\left(B_1^c(0)\right)\right]^{-1} \to 1, \qquad \text{as } \varepsilon \to 0+,$$

$$\beta_{\varepsilon}\left[\varepsilon^{\alpha\rho}\,\ell(1/\varepsilon^{\rho})\,\mu\left(B_1^c(0)\right)\right]^{-1} \to 1, \qquad \text{as } \varepsilon \to 0+. \tag{3.30}$$

Chapter 4
The Small Deviation of the Small Noise Solution

In this chapter we shall consider the solution Y^ε of the SPDE (3.11), consisting of the deterministic Chafee–Infante equation perturbed by just the small jump part ξ^ε of our Lévy process L. We show that with probability converging to 1 as $\varepsilon \to 0+$ after having entered a ball of fixed radius r^* the maximal deviation of Y^ε from the deterministic solution u on the time interval before the first big jump T_1, given by $\|Y^\varepsilon(t) - u(t)\|$ is at most of order ε to some positive power. This result is crucial for determining the asymptotic behavior of the first exit time in Chap. 5, since it basically states that exits can arise only from big jumps. Recall that the jump measure of the driving noise of the stochastic Chafee–Infante equation is symmetric and regularly varying of index $\alpha \in (0, 2)$.

First, we prove that the perturbed Chafee–Infante equation (2.12) has the same property as the unperturbed one: all solutions enter a ball in H in a uniform time provided the perturbation θ is not too big. Then, we estimate the probability that Y^ε does not stay close to the deterministic solution.

4.1 Uniformly Absorbing Ball for (2.12)

Proposition 4.1. *There exists a constant r^* and a time s_{r^*} such that for any $x \in H$ and $\theta \in D(\mathbb{R}^+; H)$ with $\sup_{t\geq 0} \|\theta(t)\| \leq 1$, the solution $v_\theta(\cdot; x)$ of (2.12) satisfies*

$$\|v_\theta(t; x) + \theta(t)\| \leq r^*, \text{ for } t \geqslant s_{r^*}.$$

Proof. We fix $\theta \in D(\mathbb{R}^+; H)$ such that $\sup_{t\geq 0} \|\theta(t)\| \leq 1$ and write $v = v_\theta$. Then (2.12) can be written in the form

$$\begin{cases} \frac{\mathrm{d}v}{\mathrm{d}t} = \frac{\partial^2 v}{\partial \zeta^2} - \lambda(v^3 + 3\theta v^2 + (3\theta^2 - 1)v + \theta^3 - \theta), \\ v(0; x) = x. \end{cases}$$

A. Debussche et al., *The Dynamics of Nonlinear Reaction-Diffusion Equations with Small Lévy Noise*, Lecture Notes in Mathematics 2085, DOI 10.1007/978-3-319-00828-8_4,

We now reproduce the estimates used in the case $\theta = 0$ (see [EFNT94, Tem92]). We multiply this equation by $|v|^k v$ and integrate with respect to the space variable to obtain after integration by part[1]:

$$\frac{1}{k+2}\frac{d}{dt}|v|_{k+2}^{k+2} + (k+1)\int_0^1 |v|^k|\nabla v|^2 d\zeta + \lambda |v|_{k+4}^{k+4}$$
$$= -\lambda \int_0^1 3\theta|v|^{k+2}v + (3\theta^2-1)|v|^{k+2} + (\theta^3-\theta)|v|^k v\, d\zeta.$$

We use Hölder inequality and $|\theta|_\infty \le \|\theta\| \le 1$ and get

$$-\int_0^1 3\theta|v|^{k+2}v d\zeta \leqslant 3|v|_{k+4}^{k+3} \leqslant \frac{1}{k+4}3^{k+4}6^{k+3} + \frac{1}{6}\frac{k+3}{k+4}|v|_{k+4}^{k+4} \leqslant 3^{k+4}6^{k+3} + \frac{1}{6}|v|_{k+4}^{k+4}.$$

Similarly:

$$-\int_0^1 (3\theta^2-1)|v|^{k+2}d\zeta \leqslant \int_0^1 |v|^{k+2}d\zeta \leqslant 6^{(k+2)/2} + \frac{1}{6}|v|_{k+4}^{k+4},$$

and

$$-\int_0^1 (\theta^3-\theta)|v|^k v\, d\zeta \leqslant \int_0^1 |v|^{k+1}d\zeta \leqslant 6^{(k+1)/3} + \frac{1}{6}|v|_{k+4}^{k+4}.$$

We deduce:

$$\frac{1}{k+2}\frac{d}{dt}|v|_{k+2}^{k+2} + (k+1)\int_0^1 |v|^k|\nabla v|^2 d\zeta + \frac{\lambda}{2}|v|_{k+4}^{k+4} \le \lambda(18)^{k+4}$$

and using again Hölder inequality

$$\frac{d}{dt}|v|_{k+2}^{k+2} + \frac{\lambda(k+2)}{2}|v|_{k+2}^{k+4} \le \lambda(18)^{k+4}(k+2).$$

Recall the following elementary result (see [EFNT94]).

Lemma 4.2. *Let $a(t)$ be a positive absolutely continuous function on $(0,\infty)$ that satisfies*

$$\frac{da}{dt} + \gamma a^p \le \delta$$

[1] As is usual when doing estimates for partial differential equations, the following computations are formal. They could be easily justified by first taking a smooth θ so that v is very regular and then approximating θ by a sequence of smooth functions.

for some $p > 1$ and δ, $\delta > 0$. Then, for $t \geqslant 0$

$$a(t) \leqslant \left(\frac{\delta}{\gamma}\right)^{1/p} + \frac{1}{(\gamma(p-1)t)^{1/p-1}}.$$

We apply this result with $a = |v|_{k+2}^{k+2}$, $p = \frac{k+4}{k+2}$ $\gamma = \frac{\lambda(k+2)}{2}$ and, since $\gamma(p-1) = \lambda$, obtain for $t \geqslant \frac{1}{\lambda}$:

$$|v(t)|_{k+2}^{k+2} \leq 2^{\frac{k+2}{k+4}} 18^{k+2} + 1.$$

Thus, recalling that $|\cdot|_\infty \leqslant \limsup_{k\to\infty} |\cdot|_k$, we deduce that

$$|v(t)|_\infty \leq 18, \quad t \geqslant \frac{1}{\lambda}.$$

We now get an estimate in H. We rewrite (2.12) in the mild form between t and $t+1$:

$$v(t+1) = S(1)v(t) + \lambda \int_t^{t+1} S(t+1-s)((v+\theta)^3 - (v+\theta))ds.$$

Therefore, thanks to the smoothing effect of the heat semigroup (2.6) and the obvious inequality $|\cdot| \leq |\cdot|_\infty$,

$$\|v(t+1)\| \leq C_1 e^{-c_0 t} |v(t)|_\infty + \lambda C_1 \int_t^{t+1} (t+1-s)^{-1/2} |(v+\theta)^3 - (v+\theta)|_\infty ds.$$

Thus, for $t \geqslant \frac{1}{\lambda}$,

$$\|v(t+1)\| \leq 18C_1 + 2\lambda C_1(19^3 + 19).$$

The result follows with $s_{r^*} = \frac{1}{\lambda} + 1$ and $r^* - 1$ equal to the right hand side above. □

Recalling our notation in Chap. 3 for the small jump part we define its stochastic convolution ξ^* by

$$\xi^*(t) = \int_0^t S(t-s)\mathrm{d}\xi^\varepsilon(s), \quad t \geqslant 0. \tag{4.1}$$

Since $Y^\varepsilon(t;x) = v_{\varepsilon\xi^*}(t;x) + \varepsilon\xi^*(t)$, we may deduce a bound on Y^ε. For $x \in H$ we define

$$\sigma_{r^*}(x,\varepsilon) := \inf\{t > 0 \mid Y^\varepsilon(t;x) \in B_{r^*}(0)\}. \tag{4.2}$$

We deduce that conditioned on small noise convolution amplitude this random time has a deterministic uniform upper bound.

Lemma 4.3. *For any $x \in H$ and $\varepsilon > 0$*

$$\sigma_{r^*}(x, \varepsilon) \leq s_{r^*},$$

on the event $\mathscr{E}_{s_{r^*}}(1) := \{\sup_{t \in [0, s_{r^*}]} \|\varepsilon\xi^*(t)\| \leq 1\}$.

Also, we give the following result which implies a sharper bound on the solution of (2.4) than the bound used in Lemma 2.2.

Lemma 4.4. *For any $x \in H$, the solution $u(t; x)$ of* (2.4) *satisfies the following bound*

$$\|u(t;x)\| \leqslant e^{-t}\|x\|^2(1+\lambda^2) + \frac{\lambda^3}{4}, \; t \geqslant 0.$$

Proof. We first multiply (2.4) by u integrate in space. We obtain after integration by parts:

$$\frac{1}{2}\frac{d}{dt}|u|^2 + \|u\|^2 + \lambda|u|_4^4 = \lambda|u|^2.$$

Thus, by Poincaré and Hölder inequality

$$\frac{1}{2}\frac{d}{dt}|u|^2 + |u|^2 + \lambda|u|_4^4 \leqslant \lambda|u|_4^2 \leqslant \lambda|u|_4^4 + \frac{\lambda}{4},$$

and by Gronwall's Lemma

$$|u(t)|^2 \leqslant e^{-2t}|x|^2 + \frac{\lambda}{4}.$$

We then multiply (2.4) by $-\Delta u$ with similar arguments:

$$\begin{aligned}\frac{1}{2}\frac{d}{dt}\|u\|^2 + |\Delta u|^2 + \lambda 3\lambda \int_0^1 u^2|\nabla u|^2 d\zeta &= -\lambda \int_0^1 u\Delta u d\zeta \\ &\leqslant \lambda|u|\,\|\Delta u| \\ &\leqslant \frac{\lambda^2}{2}|u|^2 + \frac{1}{2}|\Delta u|^2.\end{aligned}$$

Again by Poincaré inequality:

$$\frac{d}{dt}\|u\|^2 + \|u\|^2 \leqslant \lambda^2|u|^2 \leqslant \lambda^2\left(e^{-2t}|x|^2 + \frac{\lambda}{4}\right)$$

and by Gronwall Lemma

$$\|u(t)\|^2 \le \|x\|^2 e^{-t} + \lambda^2 e^{-t}|x|^2 + \frac{\lambda^3}{4} \leqslant e^{-t}\|x\|^2(1+\lambda^2) + \frac{\lambda^3}{4}. \qquad \square$$

4.2 Small Deviations of the Small Noise Solution

We define the following perturbation event, which will play a key role in Chap. 5. For r^* and s_{r^*} determined in the preceding Sect. 4.1, $x \in D^\pm$ and $\gamma, \varepsilon \in (0,1)$ we define the small deviation event

$$E_x := \mathscr{E}_{s_{r^*}}(\varepsilon^{2\gamma}) \cap \{ \sup_{s\in[\tau^*, T_1]} \|Y^\varepsilon(s;x) - u(s-\tau^*; Y^\varepsilon(\tau^*;x))\| \leqslant (1/2)\varepsilon^{2\gamma} \}, \tag{4.3}$$

where $\tau^* = T_1 \wedge \sigma_{r^*}(x,\varepsilon)$ and $\mathscr{E}_{s_{r^*}}(\varepsilon^{2\gamma}) = \{\sup_{t\in[0,s_{r^*}]} \|\varepsilon\xi^*(t)\| \leqslant \varepsilon^{2\gamma}\}$. Recall that Lemma 4.3 implies $\sigma_{r^*}(x,\varepsilon) \leqslant s_{r*}$ on E_x, which will play a crucial role in order to prove the main theorem of this section.

Proposition 4.5. *For $\alpha \in (0,2)$, there is a constant $\Gamma > 0$ such that for*

$$0 < \Theta < \frac{2-\alpha}{\alpha}, \qquad \rho \in (\frac{1}{2}, \frac{2-\alpha}{2-(1-\Theta)\alpha}) \qquad 0 < \gamma < \frac{(2-\alpha)(1-\rho) - \Theta\alpha\rho}{2(\Gamma+2)}$$

there exist constants $\vartheta = \vartheta(\Theta, \rho, \gamma, \alpha) > \alpha(1-\rho)$, $C_\vartheta > 0$ and $\varepsilon_0 > 0$, such that for all $0 < \varepsilon \leqslant \varepsilon_0$

$$\mathbb{E}\left[e^{\lambda^\pm(\varepsilon)T_2} \sup_{x\in D^\pm(\varepsilon^\gamma)} \mathbf{1}(E_x^c) \right] \leqslant C_\vartheta \varepsilon^\vartheta.$$

Recall that $\lambda \pm (\varepsilon)$ was defined in Definition 3.46 and that by (3.30) for ε small enough $\lambda^\pm(\varepsilon) < \beta_\varepsilon$.

The proof is completed in Sect. 4.4 after a series of partial results in Sects. 4.3.1 and 4.3.2. This result will be useful in Chap. 5 in the following form.

Corollary 4.6. *Let $C > 0$, and let the assumptions of Proposition 4.5 be satisfied. Then there is $\varepsilon_0 > 0$ such that for all $0 < \varepsilon \leqslant \varepsilon_0$, $\theta > -1$*

$$\mathbb{E}\left[e^{-\theta\lambda^\pm(\varepsilon)T_2} \sup_{x\in D^\pm(\varepsilon^\gamma)} \mathbf{1}(E_x^c) \right] \leqslant C \left(\frac{\beta_\varepsilon}{\beta_\varepsilon + \theta\lambda^\pm(\varepsilon)} \right)^{\frac{\lambda^\pm(\varepsilon)}{\beta_\varepsilon}}. \tag{4.4}$$

Proof. Since $\vartheta > \alpha(1-\rho)$ and C_ϑ according to the preceding Proposition 4.5, by the asymptotic properties of the functions β_ε and $\lambda^\pm(\varepsilon)$ stated in (3.30) we may conclude that there exists $\varepsilon_0 > 0$ such that for $0 < \varepsilon \leqslant \varepsilon_0$ we have

$$C_\vartheta \varepsilon^\vartheta \leqslant C\left(\frac{\beta_\varepsilon}{\beta_\varepsilon + \theta\lambda^\pm(\varepsilon)}\right)\frac{\lambda^\pm(\varepsilon)}{\beta_\varepsilon}.$$ □

4.3 Small Deviation on Deterministic Time Intervals

First we argue how to reduce a version of E_x with a deterministic time horizon $T > 0$ instead of T_1 to a set where the driving noise is small. More precisely, for $t \geqslant 0$ and $\eta > 0$, we set

$$\mathscr{E}_t(\eta) = \{\sup_{s\in[0,t]} \|\varepsilon\xi^*(s)\| \leqslant \eta\},$$

and prove that $\mathscr{E}_T(\varepsilon^{(\Gamma+2)\gamma})$ is a subset of E_x if T_1 is replaced by a finite horizon. Secondly we estimate the probability of $\mathscr{E}_T(\varepsilon^{2(\Gamma+2)})$.

4.3.1 Small Deviation with Controlled Small Noise Convolution

The aim of this subsection is the proof of Proposition 4.7, which turns out to be a consequence of the combination of a number of lemmas. We shall show that, provided the initial data is in $B_{r^*}(0)$ the deviation of the small noise mild solution from the solution of the deterministic Chafee–Infante equation u is small if the convolution of the small noise is uniformly controlled on finite deterministic time intervals.

Proposition 4.7. *There is a constant $\Gamma > 0$ such that for $0 < \alpha < 2$, $\gamma \in (0,1)$ and $\rho \in (\frac{1}{2}, 1)$ there exist $\varepsilon_0 > 0$ such that, for any $T > 0$, $0 < \varepsilon \leqslant \varepsilon_0$,*

$$\mathscr{E}_T(\varepsilon^{(\Gamma+2)\gamma}) \subseteq \bigcap_{x\in D^\pm(\varepsilon^\gamma)\cap B_{r^*}(0)} \{\sup_{s\in[0,T]} \|Y^\varepsilon(s;x) - u(s;x)\| < (1/2)\varepsilon^{2\gamma}\} \quad (4.5)$$

We set:

$$R^\varepsilon(\cdot;x) := Y^\varepsilon(\cdot;x) - u(\cdot;x) - \varepsilon\xi^*(\cdot),$$

where ξ^* is defined in Chap. 3, see (4.1).

Lemma 4.8. *For any $T_0 > 0, \tilde{\kappa} > 0$ there is a constant $\Gamma = \Gamma(\tilde{\kappa}) > 0$ such that for $\rho \in (1/2, 1)$, $K > 0$ and $\gamma > 0$ there exists $\varepsilon_0 = \varepsilon_0(K, T_0, \gamma, \tilde{\kappa}) > 0$ such that for $0 < \varepsilon \leqslant \varepsilon_0$, $x \in D^{\pm}(\varepsilon^{\gamma}) \cap B_{r^*}(0)$, and $0 \leqslant T \leqslant T_0 + \tilde{\kappa}\gamma|\ln\varepsilon|$, the remainder process $R^{\varepsilon}(\cdot; x) := Y^{\varepsilon}(\cdot; x) - u(\cdot; x) - \varepsilon\xi^*(\cdot)$ satisfies*

$$\sup_{t\in[0,T]} \|R^{\varepsilon}(t; x)\| \leqslant \frac{1}{K}\varepsilon^{2\gamma}$$

on the event $\mathscr{E}_T(\varepsilon^{(\Gamma+2)\gamma}) := \{\sup_{r\in[0,T]} ||\varepsilon\xi^*(r)|| < \varepsilon^{(\Gamma+2)\gamma}\}$.

Proof. The proof is similar to the proof of Lemma 2.17. Let $x \in D^{\pm}(\varepsilon^{\gamma}) \cap B_{r^*}(0)$. We first note that by Lemma 4.4, since $\|x\| \leqslant r^*$, we have

$$\|u(t; x)\| \leqslant R^{**} = (1 + \lambda^2)r^* + \frac{\lambda^3}{4}, \ t \geqslant 0.$$

The process $R^{\varepsilon}(\cdot; x)$ for which we note briefly R^{ε} in the sequel satisfies the equation

$$\frac{\mathrm{d}R^{\varepsilon}}{\mathrm{d}t} = \Delta R^{\varepsilon} + f(Y^{\varepsilon}) - f(u).$$

We use the integral form of this equation and use (2.5) to get, recalling that H is an algebra,

$$\begin{aligned}\|R^{\varepsilon}(t)\| &\leqslant \int_0^t e^{-c_0(t-s)}\|f(Y^{\varepsilon}) - f(u)\|ds \\ &\leqslant \lambda \int_0^t \left(\|Y^{\varepsilon}\|^2 + \|Y^{\varepsilon}\|\,\|u\| + \|u\|^2 + 1\right)\|R^{\varepsilon} + \varepsilon\xi^*\|ds.\end{aligned}$$

Choosing $\varepsilon_0 \leqslant 1$, on $\mathscr{E}_T(\varepsilon^{(\Gamma+2)\gamma})$ and for $t \leqslant \tau = \inf\{t \geqslant 0,\ \|R^{\varepsilon}\| \geqslant 1\}$, we have for $t \leqslant \tau$

$$\|Y^{\varepsilon}(t)\| \leqslant \|u(t)\| + \|R^{\varepsilon}(t)\| + \|\varepsilon\xi^*(t)\| \leqslant R^{**} + 2,$$

and

$$\begin{aligned}\|R^{\varepsilon}(t)\| &\leqslant 3\lambda(R^{**} + 2)^2 \int_0^t \|R^{\varepsilon}\|ds + 3\lambda(R^{**} + 2)^2 \int_0^t e^{-c_0(t-s)}\|\varepsilon\xi^*\|ds \\ &\leqslant 3\lambda(R^{**} + 2)^2 \int_0^t \|R^{\varepsilon}\|ds + \frac{1}{c_0}3\lambda(R^{**} + 2)^2 \sup_{t\in[0,T]} \|\varepsilon\xi^*(t)\|.\end{aligned}$$

Then, by Gronwall's Lemma, on $\mathscr{E}_T(\varepsilon^{(\Gamma+2)\gamma})$ and for $t \leqslant \tau$:

$$\|R^{\varepsilon}(t)\| \leqslant \frac{1}{c_0}3\lambda(R^{**} + 2)^2 \sup_{t\in[0,T]} \|\varepsilon\xi^*(t)\|e^{3\lambda(R^{**}+2)^2 t}.$$

Let $\tilde{\kappa} > 0$ and set $\Gamma = 3\lambda(R^{**}+2)^2\tilde{\kappa} + 1$, then if moreover $t \leqslant T_0 + \tilde{\kappa}\gamma \ln|\varepsilon|$,

$$\begin{aligned} \|R^\varepsilon(t)\| &\leqslant \frac{1}{c_0} 3\lambda(R^{**}+2)^2 e^{3\lambda(R^{**}+2)^2 t} \varepsilon^{(\Gamma+2)\gamma} \varepsilon^{-\Gamma\gamma+\gamma} \\ &\leqslant \frac{1}{c_0} 3\lambda(R^{**}+2)^2 e^{3\lambda(R^{**}+2)^2 t} \varepsilon^{2\gamma} \varepsilon^{\gamma} \leqslant \frac{1}{K} \varepsilon^{2\gamma} \end{aligned} \tag{4.6}$$

for $\varepsilon \leqslant \varepsilon_0$ and a well chosen ε_0. We may also choose ε_0 such that $\varepsilon_0^{2\gamma} \leqslant K$, we deduce that $\tau \geqslant T_0 + \tilde{\kappa}\gamma|\ln\varepsilon|$ on $\mathscr{E}_T(\varepsilon^{(\Gamma+2)\gamma})$ and thus the result. □

Note that the result stays true for any $\tilde{\Gamma} \geqslant \Gamma$.

Lemma 4.9. *Let $\phi \in \{\phi^+, \phi^-\}$, there exists $0 < \delta_0$, δ_1 and a constant $\kappa_4 > 0$ such that for all $\rho \in (1/2, 1)$, $x \in B_{\delta_0}(\phi)$, $0 < \varepsilon \leqslant 1$ and $0 \leqslant T$ we have*

$$\sup_{t\in[0,T]} \|R^\varepsilon(t;x)\| \leqslant \kappa_4 \sup_{r\in[0,T]} \|\varepsilon\xi^*(r)\|$$

on $\mathscr{E}_T(\delta_1)$.

Proof. As in Lemma 2.18, we use the stability property of ϕ and denote by Λ a positive constant such that that for all $w \in H$

$$\langle \Delta w + f'(v)w, w\rangle \leqslant -\Lambda|w|^2. \tag{4.7}$$

We again take $\eta_1 \in (0,1]$ such that for any $w \in B_{\eta_1}(\phi^\pm)$ we have

$$\|f'(w) - f'(\phi^\pm)\| \leq \frac{\Lambda}{4}, \quad \|f'(w)\| \leq 2M_1,$$

where $M_1 = \max\{\|f'(\phi^+)\|, \|f'(\phi^-)\|\}$.

By the stability of ϕ, the exists $\delta_0 > 0$ such that, for $x \in B_{\delta_0}(\phi)$, $u(t;x) \in B_{\eta_1/2}(\phi)$ for all $t \geq 0$.

Define

$$\tau = \inf\{t \geqslant 0 \mid \|R^\varepsilon(t;x)\| > \frac{\eta_1}{4}\}.$$

Then, if $\delta_1 \leqslant \frac{\eta_1}{4}$, for $t \leqslant \tau \wedge T$, $Y^\varepsilon \in B_{\eta_1}(\phi)$ on $\mathscr{E}_T(\delta_1)$. Let us write:

$$\begin{aligned} \frac{\mathrm{d}R^\varepsilon}{\mathrm{d}t} &= \Delta R^\varepsilon + (f(Y^\varepsilon) - f(u)) \\ &= \Delta R^\varepsilon + \left(\int_0^1 f'(u + \theta_1(R^\varepsilon + \varepsilon\xi^*))\,\mathrm{d}\theta_1\right)(R^\varepsilon + \varepsilon\xi^*) \\ &= \Delta R^\varepsilon + f'(v)R^\varepsilon + \left(\int_0^1 f'(u + \theta_1(R^\varepsilon + \varepsilon\xi^*)) - f'(v)\,\mathrm{d}\theta_1\right)R^\varepsilon \\ &\quad + \left(\int_0^1 f'(u + \theta_1(R^\varepsilon + \varepsilon\xi^*))\,\mathrm{d}\theta_1\right)\varepsilon\xi^*. \end{aligned} \tag{4.8}$$

After a similar operation as in Lemma 2.18, we deduce that on $\mathscr{E}_T(\delta_1)$ and for $t \leq \tau \wedge T$:

$$\frac{1}{2}\frac{\mathrm{d}}{\mathrm{d}t}|R^\varepsilon|^2 + \Lambda|R^\varepsilon|^2 \leqslant \frac{\Lambda}{4}|R^\varepsilon|^2 + 2M_1|R^\varepsilon|\,|\varepsilon\xi^*|,$$

and

$$\frac{\mathrm{d}}{\mathrm{d}t}|R^\varepsilon|^2 + \Lambda|R^\varepsilon|^2 \leqslant \frac{4M_1^2}{\Lambda}|\varepsilon\xi^*|^2.$$

Gronwall's Lemma and $R^\varepsilon(0) = 0$ imply under these conditions for times $t \leqslant \tau \wedge T$ that

$$\sup_{s\in[0,t]} |R^\varepsilon(s;x)|^2 \leqslant \frac{4M_1^2}{\Lambda^2} \sup_{s\in[0,t]} |\varepsilon\xi^*(s)|^2$$

on $\mathscr{E}_T(\delta_1)$.

We then sharpen the estimate obtained in the first part to an estimate in the $\|\cdot\|$–norm thanks to (2.6):

$$\|R^\varepsilon(t)\| \leqslant C_1 \int_0^t e^{-(c_0)(t-r)}(t-r)^{-1/2}|f(Y^\varepsilon(r)) - f(u(r))|\,\mathrm{d}r.$$

Therefore for $t \leqslant \tau \wedge T$ and on $\mathscr{E}(\delta_1)$:

$$\begin{aligned}
\|R^\varepsilon(t)\| &\leqslant 2C_1M_1 \int_0^t e^{-(c_0)(t-r)}(t-r)^{-1/2}|R^\varepsilon(r) + \varepsilon\xi^*(r)|\mathrm{d}r \\
&\leqslant 2C_1M_1(1 + \frac{2M_1}{\Lambda}) \sup_{s\in[0,t]} |\varepsilon\xi^*(s)| \int_0^t e^{-(c_0)(t-r)}(t-r)^{-1/2}\,\mathrm{d}r \\
&\leqslant \kappa_4 \sup_{s\in[0,t]} |\varepsilon\xi^*(s)|,
\end{aligned}$$

with $\kappa_4 = 2C_1M_1(1 + \frac{2M_1}{\Lambda}) \int_0^\infty e^{-c_0 t} t^{-1/2}$.

Take $\delta_1 \leqslant \kappa_4^{-1}$, then $\|R^\varepsilon(t;x)\| \leqslant 1$ on $\mathscr{E}(\delta_1)$. It follows that $\tau \geqslant T$ on $\mathscr{E}(\delta_1)$ and

$$\|R^\varepsilon(t)\| \leqslant \kappa_4 \sup_{s\in[0,T]} |\varepsilon\xi^*(s)|^2,\ t \leqslant T,$$

on $\mathscr{E}(\delta_1)$. □

We finally combine the results of the preceding two lemmas to obtain a uniform estimate for the remainder process R^ε.

Lemma 4.10. *There is a constant $\Gamma > 0$ such that for $\rho \in (1/2, 1), \gamma > 0$, there exists $\varepsilon_0 > 0$ such that for $0 < \varepsilon \leqslant \varepsilon_0$, $T > 0$, $x \in D^{\pm}(\varepsilon^\gamma) \cap B_{r^*}(0)$ on the event $\mathscr{E}_T(\varepsilon^{(\Gamma+2)\gamma})$ we have the estimate*

$$\sup_{t\in[0,T]} \|R^\varepsilon(t;x)\| \leqslant \frac{1}{4}\varepsilon^{2\gamma}.$$

Proof. Let us fix $\gamma > 0$, $T \geq 0$. Let $\kappa > 0$ and T_{rec} be given by Proposition 2.12. We use Lemma 4.8 with $\tilde{\kappa} = \kappa$ and $T_0 = T_{rec}$, this defines $\Gamma = \Gamma(\kappa)$.

To lighten notations, we set:

$$T_\varepsilon^* = T_{rec} + \kappa\gamma|\ln\varepsilon|.$$

By Lemma 4.8, if ε_0 is sufficiently small we have for $t \leqslant T_\varepsilon^*$

$$\sup_{t\in[0,T]} \|R^\varepsilon(t;x)\| \leq \frac{1}{K}\varepsilon^{2\gamma}$$

on $\mathscr{E}_T(\varepsilon^{(\Gamma+2)\gamma})$ where K is a fixed constant to be determined below. This implies the result for $t \leqslant T_\varepsilon^*$ provided $K \geqslant 4$.

Let us consider now the case $t \geqslant T_\varepsilon^*$. We first observe that by Proposition 2.12

$$\|u(T_\varepsilon^*;x) - \phi)\| \leqslant \frac{1}{2}\varepsilon^{2\gamma},$$

where $\phi \in \{\phi^+, \phi^-\}$ is such that $u(t;x) \to \phi$ as $t \to \infty$. It follows

$$\begin{aligned}\|Y^\varepsilon(T_\varepsilon^*;x) - \phi\| &\leqslant \|u(T_\varepsilon^*;x) - \phi\| + \|R^\varepsilon(T_\varepsilon^*;x)\| + \|\varepsilon\xi^*(T_\varepsilon^*)\| \\ &\leqslant \frac{1}{2}\varepsilon^{2\gamma} + \frac{1}{K}\varepsilon^{2\gamma} + \varepsilon^{(\Gamma+2)\gamma} \leqslant \varepsilon^{2\gamma}\end{aligned}$$

on $\mathscr{E}(\varepsilon^{(\Gamma+2)\gamma})$ provided $\varepsilon_0^{\Gamma\gamma} \leqslant \frac{1}{4}$.

Using similar arguments as in Lemma 2.2 and the fact that, for $x \in B_{\delta_0}(\phi)$, $u(t;x) \in B_{\delta_1}(\phi)$, we easily find a constant L_0 such that

$$\|u(t;x) - u(t;y)\| \leqslant L_0\|x-y\|,\ t \geqslant 0,\ x, y \in B_{\delta_0}(\phi).$$

Then, if $\varepsilon_0^{2\gamma} \leqslant \delta_0$, we have on $\mathscr{E}(\varepsilon^{(\Gamma+2)\gamma})$

$$\begin{aligned}\|u(t;x) - u(t-T_\varepsilon^*;Y^\varepsilon(T_\varepsilon^*,x))\| &= \|u(t-T_\varepsilon^*;u(T_\varepsilon^*,x)) - u(t-T_\varepsilon^*;Y^\varepsilon(T_\varepsilon^*,x))\| \\ &\leqslant L_0\|u(T_\varepsilon^*,x)) - Y^\varepsilon(T_\varepsilon^*,x))\| \\ &\leqslant L_0\left[\|R^\varepsilon(T_\varepsilon^*;x)\| + \|\varepsilon\xi^*(T_\varepsilon^*)\|\right] \\ &\leqslant L_0\left(\frac{1}{K}\varepsilon^{2\gamma} + \varepsilon^{(\Gamma+2)\gamma}\right).\end{aligned}$$

We deduce, still on $\mathscr{E}(\varepsilon^{(\Gamma+2)\gamma})$,

$$\begin{aligned}\|R^{\varepsilon}(t;x)\| &= \|Y^{\varepsilon}(t-T_{\varepsilon}^{*},T_{\varepsilon}^{*};Y^{\varepsilon}(T_{\varepsilon}^{*},x)) - u(t-T_{\varepsilon}^{*};u(T_{\varepsilon}^{*},x)) + \varepsilon\xi^{*}(t)\| \\ &\leqslant \|Y^{\varepsilon}(t-T_{\varepsilon}^{*},T_{\varepsilon}^{*};Y^{\varepsilon}(T_{\varepsilon}^{*},x)) - u(t-T_{\varepsilon}^{*};Y^{\varepsilon}(T_{\varepsilon}^{*},x)) \\ &\quad + \varepsilon\xi^{*}(t)\| + L_0\left(\frac{1}{K}\varepsilon^{2\gamma} + \varepsilon^{(\Gamma+2)\gamma}\right),\end{aligned}$$

where $Y^{\varepsilon}(t,s;y)$ denotes the small noise solution, *i.e.* the solution of (3.11), at time t with an initial data equal to y at time s.

Clearly, the statement of Lemma 4.9 extends to our situation where the time is translated by T_{ε}^{*} and we obtain that, if $\varepsilon_0^{(\Gamma+2)\gamma} \leqslant \delta_1$, we have

$$\|R^{\varepsilon}(t;x)\| \leqslant \kappa_4\varepsilon^{(\Gamma+2)\gamma} + \varepsilon^{(\Gamma+2)\gamma} + L_0\left(\frac{1}{K}\varepsilon^{2\gamma} + \varepsilon^{(\Gamma+2)\gamma}\right)$$

on $\mathscr{E}(\varepsilon^{(\Gamma+2)\gamma})$. We conclude the proof by taking $K=8L_0$ and $(\kappa_4+L_0+1)\varepsilon_0^{\Gamma\gamma} \leqslant \frac{1}{8}$.
□

Proof (of Proposition 4.7).
By Lemma 4.10, we find $\Gamma > 0$ such that given $\rho \in (1/2,1), \gamma > 0$, there exists $\varepsilon_0 > 0$ such that for $0 < \varepsilon \leqslant \varepsilon_0, T > 0$, and all $x \in D^{\pm}(\varepsilon^{\gamma}) \cap B_{r^*}(0)$ we have by definition of $\mathscr{E}_T(\varepsilon^{(\Gamma+2)\gamma})$

$$\begin{aligned}&\{\sup_{t\in[0,T]} \|Y^{\varepsilon}(t;x) - u(t;x)\| \geqslant (1/2)\varepsilon^{2\gamma}\} \\ &\quad = \{\sup_{t\in[0,T]} \|R^{\varepsilon}(t;x) + \varepsilon\xi^{*}(t)\| \geqslant (1/2)\varepsilon^{2\gamma}\} \\ &\quad \subseteq \{\sup_{t\in[0,T]} \|R^{\varepsilon}(t;x)\| \geqslant (1/4)\varepsilon^{2\gamma}\} \cup \{\sup_{t\in[0,T]} \|\varepsilon\xi^{*}(t)\| \geqslant (1/4)\varepsilon^{2\gamma}\} \\ &\quad \subseteq \{\sup_{t\in[0,T]} \|\varepsilon\xi^{*}(t)\| \geqslant \varepsilon^{(\Gamma+2)\gamma}\} \cup \{\sup_{t\in[0,T]} \|\varepsilon\xi^{*}(t)\| \geqslant (1/4)\varepsilon^{2\gamma}\} \\ &\quad \subseteq \{\sup_{t\in[0,T]} \|\varepsilon\xi^{*}(t)\| \geqslant \varepsilon^{(\Gamma+2)\gamma}\} = \mathscr{E}_T^{c}(\varepsilon^{(\Gamma+2)\gamma}).\end{aligned} \tag{4.9}$$

□

4.3.2 Control of the Small Noise Convolution

In this subsection we shall deal with estimating the convolution of small noise with the semigroup of the heat equation on the unit interval, uniformly on finite deterministic time intervals. Note that in the statement of the following Lemma neither ρ nor γ are restricted within their ranges.

Lemma 4.11. *For any $\rho \in (0,1)$, $p > 0$ and $0 < \Theta < 1$, there exist $\kappa_5(\rho, p, \Theta) > 0$ and $\varepsilon_0 > 0$ such that for $0 < \varepsilon \leqslant \varepsilon_0$ and $T \geqslant 0$*

$$\mathbb{P}\Big(\sup_{t\in[0,T]} \|\varepsilon\xi_t^*\| \geqslant \varepsilon^p\Big) \leqslant \kappa_5\, T\, \varepsilon^{2-2p-(2-(1-\Theta)\alpha)\rho}.$$

Proof. 1. We first show that there exists $K_1 > 0$ such that for any $\rho \in (0,1)$, $p > 0$, $\varepsilon > 0$

$$\mathbb{P}\Big(\sup_{t\in[0,T]} \|\varepsilon\xi^*(t)\| > \varepsilon^p\Big) \leqslant K_1\, \varepsilon^{-2(p-1)}\, T \Big(\int_{\{0<\|y\|\leqslant \frac{1}{\varepsilon^\rho}\}} \|y\|^2\, \nu(\mathrm{d}y)\Big). \tag{4.10}$$

We start by applying Kolmogorov's inequality, to get

$$\mathbb{P}\Big(\sup_{t\in[0,T]} \|\varepsilon\xi^*(t)\| > \varepsilon^p\Big) \leqslant (\varepsilon^{p-1})^{-2}\, \mathbb{E}\Big[\sup_{t\in[0,T]} \|\xi^*(t)\|^2\Big].$$

Now consider the stochastic convolution equation

$$\mathrm{d}\xi^* = \Delta\xi^*\,\mathrm{d}t + \mathrm{d}\xi^\varepsilon, \qquad \xi^*(0) = 0.$$

For $t \geqslant 0$ we denote by $\Delta_t X = X(t) - X(t-)$ the jump of a càdlàg process X at time t, and remark that by definition $\Delta_t\xi^* = \Delta_t\xi^\varepsilon$. By Itô's formula we can write for $T \geqslant 0$

$$\begin{aligned}
\|\xi^*(T)\|^2 &= 2\int_0^T \langle \xi^*(s-), \mathrm{d}\xi^*(s)\rangle_H \\
&\quad + \sum_{s\leqslant T} \big(\|\xi^*(s)\|^2 - \|\xi^*(s-)\|^2 - 2\langle \xi^*(s-), \Delta_s\xi^\varepsilon\rangle_H\big) \\
&= 2\int_0^T \langle \xi^*(s-), \Delta\xi^*(s-)\rangle_H\,\mathrm{d}s + 2\int_0^T \langle \xi^*(s-), \mathrm{d}\xi^\varepsilon(s)\rangle_H \\
&\quad + \sum_{s\leqslant T} \big(\|\xi^*(s)\|^2 - \|\xi^*(s-)\|^2 - 2\langle \xi^*(s-), \Delta_s\xi^\varepsilon\rangle_H\big).
\end{aligned}$$

By the non-positivity of $\int_0^T \langle \xi^*(s-), \Delta\xi^\varepsilon(s-)\rangle_H\,\mathrm{d}s$ we may continue to estimate

$$\begin{aligned}
\|\xi^*(T)\|^2 &\leqslant 2\int_0^T \langle \xi^*(s-), \mathrm{d}\xi^\varepsilon(s)\rangle_H \\
&\quad + \sum_{s\leqslant T} \big(\|\xi^*(s)\|^2 - \|\xi^*(s-)\|^2 - 2\langle \xi^*(s-), \Delta_s\xi^\varepsilon\rangle_H\big).
\end{aligned}$$

Note that for $s \leqslant T$

$$\|\xi^*(s)\|^2 - \|\xi^*(s-)\|^2 - 2\langle \xi^*(s-), \Delta_s \xi^\varepsilon \rangle_H = \|\Delta_s \xi^*\|^2 = \|\Delta_s \xi^\varepsilon\|^2,$$

and therefore

$$\|\xi^*(T)\|^2 \leqslant \int_0^T \langle \xi^*(s-), \mathrm{d}\xi^*(s) \rangle_H + \sum_{s \leqslant T} \|\Delta_s \xi^\varepsilon\|^2.$$

For $t \geqslant 0$ let us denote by $[[X]]_t$ the quadratic variation of a process X on $[0, t]$. Then Burkholder's inequality yields a universal constant $C_2 > 0$ such that by Young's inequality for any $a > 0$ we have

$$\begin{aligned}
&\mathbb{E}\Big[\sup_{s \in [0,T]} \|\xi^*(s)\|^2\Big] \\
&\leqslant 2\,\mathbb{E}\Big[\sup_{s \in [0,T]} \Big| \int_0^s \langle \xi^*(r-), \mathrm{d}\xi^\varepsilon(r) \rangle_H \Big|\Big] + \mathbb{E}\Big[\sum_{r \leqslant T} \|\Delta_r \xi^*\|^2\Big] \\
&\leqslant 2C_2\,\mathbb{E}\Big[[[\int_0^\cdot \langle \xi^*(r-), \mathrm{d}\xi^\varepsilon(r) \rangle_H]]_T^{1/2}\Big] + T \int_{\{0<\|y\| \leqslant 1/\varepsilon^\rho\}} \|y\|^2\, \mathrm{d}\nu(\mathrm{d}y) \\
&= 2C_2\,\mathbb{E}\Big[\Big(\sup_{s \in [0,T]} \|\xi^*(s)\|^2 \int_0^T \mathrm{d}[[\xi^*]](s)\Big)^{1/2}\Big] + T \int_{\{0<\|y\| \leqslant 1/\varepsilon^\rho\}} \|y\|^2\, \mathrm{d}\nu(\mathrm{d}y) \\
&\leqslant 2C_2 \Big(a\,\mathbb{E}\Big[\sup_{s \in [0,T]} \|\xi^*(s)\|^2\Big] + \frac{1}{4a}\mathbb{E}\Big[\int_0^T \mathrm{d}[[\xi^*]](s)\Big]\Big) \\
&\quad + T \int_{\{0<\|y\| \leqslant 1/\varepsilon^\rho\}} \|y\|^2\, \mathrm{d}\nu(\mathrm{d}y) \\
&= 2aC_2\,\mathbb{E}\Big[\sup_{s \in [0,T]} \|\xi^*(s)\|^2\Big] + \frac{C_2\, T}{2a} \int_{\{0<\|y\| \leqslant \frac{1}{\varepsilon^\rho}\}} \|y\|^2\, \nu(\mathrm{d}y) \\
&\quad + T \int_{\{0<\|y\| \leqslant 1/\varepsilon^\rho\}} \|y\|^2\, \mathrm{d}\nu(\mathrm{d}y).
\end{aligned}$$

Choosing now $a = 1/(4C_2)$ we obtain

$$\mathbb{E}\left[\sup_{s\in[0,T]} \|\xi^*(s)\|^2\right] \leqslant \left(4C_2^2 + 2\right) T \int_{\{0<\|y\|\leqslant\frac{1}{\varepsilon^\rho}\}} \|y\|^2 \, \nu(\mathrm{d}y).$$

Now take $K_1 = 4C_2^2 + 2$ to finish our argument.

2. In the second part of the proof it remains to determine the asymptotic behavior of the last factor for small $0 < \varepsilon < 1$. We first write

$$\int_{\{0<\|y\|\leqslant\frac{1}{\varepsilon^\rho}\}} \|y\|^2 \, \nu(\mathrm{d}y) \leqslant \int_{\{0<\|y\|\leqslant 1\}} \|y\|^2 \, \nu(\mathrm{d}y) + \int_{\{1<\|y\|\leqslant\frac{1}{\varepsilon^\rho}\}} \|y\|^2 \, \nu(\mathrm{d}y),$$

and remark that by part 1 of the proof it remains to estimate the asymptotic behavior of the function

$$\varepsilon \mapsto \int_{\{1<\|y\|\leqslant\frac{1}{\varepsilon^\rho}\}} \|y\|^2 \, \nu(\mathrm{d}y)$$

for small $0 < \varepsilon < 1$. To do this, we use the regular variation of $t \mapsto \nu(tB_1^c(0)) = t^{-\alpha} \, \ell(t) \, \mu(B_1^c(0))$ with a slowly varying function ℓ and limiting measure μ (see Definition 3.44). We also use Proposition 3.43 which implies that for any slowly varying function ℓ and $1>\Theta>0$ there exists $K_2 > 0$, such that $\ell(t) \leqslant K_2 + t^{\Theta\alpha}$. This results in the following chain of inequalities

$$\begin{aligned}
\int_{\{1<\|y\|\leqslant\frac{1}{\varepsilon^\rho}\}} \|y\|^2 \, \nu(\mathrm{d}y) &= \int_H \int_0^{\|y\|} \mathbf{1}\{1 < \|y\| \leqslant \frac{1}{\varepsilon^\rho}\} 2t \, \mathrm{d}t \, \nu(\mathrm{d}y) \\
&\leqslant 2 \int_0^{\frac{1}{\varepsilon^\rho}} t \, \nu\left((1 \vee t) B_1^c(0)\right) \, \mathrm{d}t \\
&\leqslant 1 + 2\mu(B_1^c(0)) \int_1^{\frac{1}{\varepsilon^\rho}} t^{1-\alpha} \, \left(K_2 + t^{\Theta\alpha}\right) \, \mathrm{d}t \\
&= K_3 + \frac{2\, K_2 \mu(B_1^c(0))}{2-\alpha} \varepsilon^{-\rho(2-\alpha)} \\
&\quad + \frac{2\mu(B_1^c(0))}{2-(1-\Theta)\alpha} \varepsilon^{-\rho(2-(1-\Theta)\alpha)},
\end{aligned}$$

with another constant K_3. Therefore there exists $K_4 > 0$ such that for $\varepsilon > 0$ sufficiently small

$$\int\limits_{\{1<\|y\|\leqslant\frac{1}{\varepsilon^\rho}\}} \|y\|^2 \, \nu(\mathrm{d}y) \leqslant K_4 \varepsilon^{-\rho(2-(1-\Theta)\alpha)}.$$

Inserting this into inequality (4.10) we obtain the desired result. □

4.4 Small Deviation before the First Large Jump

In this section we apply the results from the previous one to finally obtain the small deviation estimate on the stochastic interval between 0 and the first big jump T_1. Since we know the law of T_1, an integration of the estimate just obtained is necessary. This will complete the proof of Proposition 4.5 for the time interval $[0, T_1]$.

Proof (of Proposition 4.5). We use the inequality derived in the preceding subsection, as well as the asymptotic behavior of the large jump rate β_ε given in (3.30). Lemma 4.3 provides the existence of the radius r^* and the deterministic upper bound $s_{r^*} > 0$ such that for all $\varepsilon > 0$

$$\mathscr{E}_{s_{r^*}}(\varepsilon^{(\Gamma+2)\gamma}) \subseteq \mathscr{E}_{s_{r^*}}(1) \subseteq \{\sigma_{r^*}(x,\varepsilon) \leqslant s_{r^*}\} = \{\exists\, t \in [0, s_{r^*}] : Y^\varepsilon(t;x) \in B_{r^*}(0)\}.$$

Further recall that by (4.11) there is $\Gamma > 0$ and $\varepsilon_0 > 0$ such that for all $0 < \varepsilon \leqslant \varepsilon_0, T > 0$, and $x \in D^\pm(\varepsilon^\gamma) \cap B_{r^*}(0)$

$$\{\sup_{t\in[0,T]} \|\varepsilon\xi^*(t)\| < \varepsilon^{(\Gamma+2)\gamma}\} \subseteq \{\sup_{t\in[0,T]} \|Y^\varepsilon(t;x) - u(t;x)\| < (1/2)\varepsilon^{2\gamma}\}.$$

We denote by $\theta_t(x)$ the (random) time shift by t on a given path of $Y^\varepsilon(\cdot,x)(\omega)$, $\omega \in \Omega$, $x \in H$ defined by

$$Y^\varepsilon(t+s,x)(\omega) = Y^\varepsilon(t, Y^\varepsilon(s,x)(\omega))(\omega) = Y^\varepsilon(t,\cdot)(\omega) \circ \theta_s(x)(\omega).$$

Due to the strong Markov property of Y^ε we know that for $t \geqslant 0$ and $x \in H$

$$Y^\varepsilon(t,x) \stackrel{d}{=} Y^\varepsilon(t,\cdot) \circ \theta_{\sigma_{r^*}(x,\varepsilon)}(x) \qquad \mathbb{P}(\cdot \mid \mathscr{E}_{s_{r^*}}(\varepsilon^{(\Gamma+2)\gamma}))\text{-a.s.}$$

Since $Y^\varepsilon(\sigma_{r^*}(x,\varepsilon), y) \in B_{r^*}(0)$ for any $x \in H$ by definition we obtain with the help of Proposition 4.7 for any $T > 0$ that for $x \in D(\varepsilon^\gamma)$

$$\mathbb{P}\Big(\Big[\mathscr{E}_{s_r*}(\varepsilon^{2\gamma}) \cap \{ \sup_{t\in[\sigma^*,T])} \|Y^\varepsilon(t,x) - u(t-\sigma_{r^*}(x,\varepsilon), Y^\varepsilon(\sigma_{r^*}(x,\varepsilon))\| < \varepsilon^{2\gamma}\}\Big]^c\Big)$$

$$\leqslant \mathbb{P}\Big(\Big[\mathscr{E}_{s_r*}(\varepsilon^{2\gamma}) \cap \bigcap_{y\in D^\pm(\varepsilon^\gamma)\cap B_{r^*}(0)} \{ \sup_{t\in[0,T]} \|Y^\varepsilon(t,x) - u(t,x)\| < \varepsilon^{2\gamma}\}\Big]^c\Big)$$

$$\leqslant \mathbb{P}(\mathscr{E}_T^c(\varepsilon^{(\Gamma+2)\gamma})), \tag{4.11}$$

where $\sigma^* = \sigma_{r^*}(y,\varepsilon) \wedge T$. By the interlacing structure of X^ε the waiting times t_1, t_2 and the event $\{\sup_{t\in[0,t]} \|\varepsilon\xi^*(t)\| \geqslant \varepsilon^{(\Gamma+2)\gamma}\}$ are independent for any $t > 0$. Exploiting inequality (4.11) and Lemma 4.11 we conclude that we can find $\Gamma > 0$ such that for given $\rho \in (1/2, 1), \gamma > 0$, and $0 < \Theta < 1$ there exist $C > 0$ and $\varepsilon_0 > 0$ such that for $0 < \varepsilon \leqslant \varepsilon_0$

$$\mathbb{E}\Big[\sup_{x\in D^\pm(\varepsilon^\gamma)} e^{\lambda(\varepsilon)T_2}\mathbf{1}(E_x^c)\Big]$$

$$\leqslant \int_0^\infty\int_0^\infty \mathbb{P}\Big(\sup_{t\in[0,t]} \|\varepsilon\xi^*(t)\| \geqslant \varepsilon^{(\Gamma+2)\gamma}\Big)\beta_\varepsilon^2 e^{(\lambda(\varepsilon)-\beta_\varepsilon)t} e^{(\lambda(\varepsilon)-\beta_\varepsilon)s}\, \mathrm{d}t\mathrm{d}s$$

$$\leqslant C\ \varepsilon^{2-2(\Gamma+2)\gamma-(2-(1-\Theta)\alpha)\rho} \left(\frac{\beta_\varepsilon}{\beta_\varepsilon - \lambda(\varepsilon)}\right)^2 \int_0^\infty t(\lambda(\varepsilon) - \beta_\varepsilon)e^{-(\lambda(\varepsilon)-\beta_\varepsilon)t}\, \mathrm{d}t$$

$$= C\ \varepsilon^{2-2(\Gamma+2)\gamma-(2-(1-\Theta)\alpha)\rho} \left(\frac{\beta_\varepsilon}{\beta_\varepsilon - \lambda(\varepsilon)}\right)^2 \frac{1}{\beta_\varepsilon - \lambda(\varepsilon)}$$

$$\leqslant C'\ \varepsilon^{2-2(\Gamma+2)\gamma-(2-(1-\Theta)\alpha)\rho-\alpha\rho},$$

for a constant C', where $\sigma^* = \sigma_{r^*}(y,\varepsilon) \wedge t$. Let now

$$\vartheta = 2 - 2(\Gamma+2)\gamma - (2-(1-\Theta)\alpha)\rho - \alpha\rho.$$

Upon setting $C_\vartheta = C'$ it remains to check the conditions under which $\vartheta > \alpha(1-\rho)$. We have

$$\begin{aligned}\vartheta - \alpha(1-\rho) &= 2 - 2(\Gamma+2)\gamma - (2-(1-\Theta)\alpha)\rho - \alpha\rho - \alpha(1-\rho)\\ &= 2 - \alpha - 2(\Gamma+2)\gamma - (2-(1-\Theta)\alpha)\rho\\ &= \underbrace{(2-\alpha)(1-\rho)}_{>0} - \Theta\alpha\rho - 2(\Gamma+2)\gamma > 0\end{aligned}$$

if and only if

$$0 < \gamma < \frac{(2-\alpha)(1-\rho) - \Theta\alpha\rho}{2(\Gamma + 2)}.$$

The right-hand side of the last inequality is positive if

$$(2-\alpha)(1-\rho) - \Theta\alpha\rho > 0 \quad \text{and thus iff} \quad \rho < \frac{2-\alpha}{2-(1-\Theta)\alpha}.$$

Since $\rho > 1/2$, the last inequality forces us to restrict $\Theta > 0$ to fulfill

$$\frac{1}{2} < \frac{2-\alpha}{2-(1-\Theta)\alpha} \quad \text{which is equivalent to} \quad \Theta < \frac{2-\alpha}{\alpha}.$$

Under these assumptions, identical to the ones formulated in the statement of Proposition 4.5, we have $\vartheta > \alpha(1-\rho)$. This completes the proof. □

Chapter 5
Asymptotic Exit Times

We shall now use the small deviations estimates of Chap. 4 in order to give a precise account of the exit times of the system described by our Chafee–Infante equation with small Lévy noise in H of the reduced domains of attraction of the stable states $\phi^{\pm}$ defined in Chap. 2. Our main line of reasoning will be based on the splitting of small and large jumps proposed there. In fact, the Chafee–Infante equation perturbed by small jumps being subject to only small deviations from the solution of the deterministic system before the first big jump, as shown in Chap. 4, and the time needed for relaxation in a small neighborhood of $\phi^{\pm}$ being only of logarithmic order in ε, exits will happen at times of big jumps that are big enough to leave the reduced domains of attraction. To characterize the asymptotic law of the exit time, we shall compute the asymptotics of its Laplace transform. Making these heuristic arguments mathematically rigorous will be the main task of this chapter.

5.1 Preparations: Event Estimates and Hypotheses on the Lévy Measure

The afore mentioned strategy is based on the combination of two steps. In a first step we estimate events that are related to the exit of the solution X^{ε} from $\tilde{D}(\varepsilon^{\gamma})$ with the help of events, whose probability can be estimated in a second step with the help of natural non-degeneracy hypotheses on the Lévy measure. Section 5.1.1 derives exactly those event estimates, while Sect. 5.1.2 states the mentioned hypotheses (H.1) and (H.2), which are shown to be suitable for the mentioned step two of the reasoning.

A. Debussche et al., *The Dynamics of Nonlinear Reaction-Diffusion Equations with Small Lévy Noise*, Lecture Notes in Mathematics 2085, DOI 10.1007/978-3-319-00828-8_5,

5.1.1 Estimates of Exit Events by Large Jump and Perturbation Events

In this subsection we shall exploit the strong Markov property of our process X^ε to rigorously define events that are capable of capturing the successive big jumps linked by periods of relaxation during which only small deviations from the deterministic solutions are possible. The strong Markov property allows us to represent X^ε recursively in the following way. Recall the notation used for the big jump compound Poisson part of our Lévy noise process from Chap. 3, and denote the shift by time t on the space of trajectories by $\theta_t, t \geqslant 0$. For any $k \in \mathbb{N}, t \in [0, t_k]$, $x \in H$ we have

$$X^\varepsilon(t + T_{k-1}; x) = Y^\varepsilon(t; X^\varepsilon(T_{k-1}; x)) \circ \theta_{T_{k-1}} + \varepsilon W_k \mathbf{1}\{t = t_k\}. \tag{5.1}$$

Also recall for $\varepsilon, \gamma \in (0, 1)$ the construction of the reduced domains of attraction $D^\pm(\varepsilon^\gamma) \supset \tilde{D}^\pm(\varepsilon^\gamma) \supset \hat{D}^\pm(\varepsilon^\gamma) \supset D^\pm(\varepsilon^\gamma, \varepsilon^{2\gamma}, \varepsilon^{2\gamma}, \varepsilon^{2\gamma})$ and their shifted versions $D_0^\pm(\varepsilon^\gamma)$, $\tilde{D}_0^\pm(\varepsilon^\gamma)$, $\hat{D}_0^\pm(\varepsilon^\gamma)$, $D_0^\pm(\varepsilon^\gamma, \varepsilon^{2\gamma}, \varepsilon^{2\gamma}, \varepsilon^{2\gamma})$ in Definition 2.9. For simplicity of notation we abbreviate

$$D_0^*(\varepsilon^\gamma) := \left(D_0^\pm \setminus \hat{D}_0^\pm(\varepsilon^\gamma)\right) + B_{\varepsilon^{2\gamma}}(0).$$

We introduce (and recall) for $\varepsilon > 0$ and $x \in \tilde{D}^\pm(\varepsilon^\gamma)$ the major events

$$\begin{aligned}
A_x &:= \{Y^\varepsilon(s; x) \in \tilde{D}^\pm(\varepsilon^\gamma),\ s \in [0, T_1] \text{ and } Y^\varepsilon(T_1; x) + \varepsilon W_1 \in \tilde{D}^\pm(\varepsilon^\gamma)\},\\
B_x &:= \{Y^\varepsilon(s; x) \in \tilde{D}^\pm(\varepsilon^\gamma), s \in [0, T_1] \text{ and } Y^\varepsilon(T_1; x) + \varepsilon W_1 \notin \tilde{D}^\pm(\varepsilon^\gamma)\},\\
A_x^- &:= \{Y^\varepsilon(s; x) \in \tilde{D}^\pm(\varepsilon^\gamma), s \in [0, T_1] \text{ and } Y^\varepsilon(T_1; x) + \varepsilon W_1 \in \hat{D}^\pm(\varepsilon^\gamma)\},\\
C_x &:= \{Y^\varepsilon(s; x) \in \tilde{D}^\pm(\varepsilon^\gamma), s \in [0, T_1] \text{ and } Y^\varepsilon(T_1; x) + \varepsilon W_1 \in \tilde{D}^\pm(\varepsilon^\gamma) \setminus \hat{D}^\pm(\varepsilon^\gamma)\},\\
A^\diamond &:= \{\varepsilon W_1 \in D_0^\pm\},\\
B^\diamond &:= \{\varepsilon W_1 \notin D_0^\pm\},\\
E_x &= \{ \sup_{t \in [0, \tau^*]} \|\varepsilon \xi^*(t)\| \leqslant \varepsilon^{2\gamma},\ \exists\, t \in [0, \tau^*]:\ Y^\varepsilon(t; x) \in B_{r^*}(0) \quad \text{and}\\
&\qquad \sup_{s \in [\tau^*, T_1]} \|Y^\varepsilon(s; x) - u(s - \tau^*; Y^\varepsilon(\tau^*; x))\| \leqslant (1/2)\varepsilon^{2\gamma} \},\\
&\qquad \text{where } \tau^* = T_1 \wedge s_{r^*}.
\end{aligned} \tag{5.2}$$

In the following two lemmas we estimate these events connecting the behaviour of X^ε in the domains of type $\tilde{D}^\pm(\varepsilon^\gamma)$ with the large jumps η^ε in the reshifted domains of type $\tilde{D}_0^\pm(\varepsilon^\gamma)$.

Lemma 5.1 (Partial estimates of the major events).
Let $T_{rec}, \kappa > 0$ given by Proposition 2.12 and $r^*, s_{r^*} > 0$ by Proposition 4.1. Then for $\rho \in (\frac{1}{2}, 1)$, $\gamma \in (0, 1-\rho)$ there exists $\varepsilon_0 > 0$ so that the following inequalities are satisfied for all $0 < \varepsilon \leqslant \varepsilon_0$ and $x \in \tilde{D}^{\pm}(\varepsilon^\gamma)$

$$i)\ \mathbf{1}(A_x)\mathbf{1}(E_x)\mathbf{1}\{T_1 \geqslant s_{r^*} + T_{rec} + \kappa\gamma|\ln\varepsilon|\} \leqslant \mathbf{1}\{\varepsilon W_1 \in D_0^{\mp}\}, \tag{5.3}$$

$$ii)\ \mathbf{1}(B_x)\mathbf{1}(E_x)\mathbf{1}\{T_1 \geqslant s_{r^*} + T_{rec} + \kappa\gamma|\ln\varepsilon|\} \leqslant \mathbf{1}\{\varepsilon W_1 \notin \hat{D}_0^{\pm}(\varepsilon^\gamma)\}, \tag{5.4}$$

$$iii)\ \mathbf{1}(C_x)\mathbf{1}(E_x)\mathbf{1}\{T_1 \geqslant s_{r^*} + T_{rec} + \kappa\gamma|\ln\varepsilon|\} \leqslant \mathbf{1}\{\varepsilon W_1 \in D_0^{*}(\varepsilon^\gamma)\}. \tag{5.5}$$

Additionally, we have

$$iv)\ \mathbf{1}(B_x)\mathbf{1}(E_x)\mathbf{1}\{\|\varepsilon W_1\| < (1/2)\varepsilon^{2\gamma}\}\mathbf{1}\{T_1 \geqslant s_{r^*} + T_{rec} + \kappa\gamma|\ln\varepsilon|\} = 0, \tag{5.6}$$

$$v)\ \mathbf{1}(C_x)\mathbf{1}(E_x)\mathbf{1}\{\|\varepsilon W_1\| < (1/2)\varepsilon^{2\gamma}\}\mathbf{1}\{T_1 \geqslant s_{r^*} + T_{rec} + \kappa\gamma|\ln\varepsilon|\} = 0. \tag{5.7}$$

In the opposite sense follows for $x \in \hat{D}^{\pm}(\varepsilon^\gamma)$

$$vi)\ \mathbf{1}(E_x)\mathbf{1}\{T_1 \geqslant s_{r^*} + T_{rec} + \kappa\gamma|\ln\varepsilon|\}\mathbf{1}\{\varepsilon W_1 \notin D_0^{\mp}\} \leqslant \mathbf{1}(B_x), \tag{5.8}$$

$$vii)\ \mathbf{1}(E_x)\mathbf{1}\{T_1 \geqslant s_{r^*} + T_{rec} + \kappa\gamma|\ln\varepsilon|\}\mathbf{1}\{\varepsilon W_1 \in D_0^{\mp}(\varepsilon^\gamma, \varepsilon^{2\gamma}, \varepsilon^{2\gamma}, \varepsilon^{2\gamma})\} \leqslant \mathbf{1}(A_x^{\mp}). \tag{5.9}$$

Proof. We fix $\gamma \in (0, 1)$.

1. We show that for $\varepsilon > 0$ small enough and $x \in \tilde{D}^{\pm}(\varepsilon^\gamma)$ the event $E_x \cap \{T_1 - s_{r^*} \geqslant T_{rec} + \kappa\gamma|\ln\varepsilon|\}$ implies that

$$Y^\varepsilon(T_1; x) \in B_{\varepsilon^{2\gamma}}(\phi^{\pm}). \tag{5.10}$$

Definition 2.9 states for $x \in \tilde{D}^{\pm}(\varepsilon^\gamma)$ and $\theta \in L^\infty(0, \infty; H)$ with $\sup_{t \geqslant 0}\|\theta(t)\| \leqslant \varepsilon^{2\gamma}$ that

$$v_\theta(t; x) + \theta(t) \in D^{\pm}(\varepsilon^\gamma) \text{ for all } t \geqslant 0.$$

Hence, the event $\{\sup_{t \in [0, T_1 \wedge s_{r^*}]}\|\varepsilon\xi^*(t)\| \leqslant \varepsilon^{2\gamma}\}$, which by Definition 5.2 is a subset of E_x, implies

$$Y^\varepsilon(t; x) = v_{\varepsilon\xi^*}(t; x) + \varepsilon\xi^*(t) \in D^{\pm}(\varepsilon^\gamma) \text{ for all } t \in [0, T_1 \wedge s_{r^*}]. \tag{5.11}$$

We choose $\varepsilon > 0$ small enough such that $T_{rec} + \kappa\gamma|\ln\varepsilon| \geqslant s_{r^*}$. Combining inclusion (5.11) and Proposition 2.12 the event $\{T_1 - s_{r^*} \geqslant T_{rec} + \kappa\gamma|\ln\varepsilon|\}$ implies then

$$u(T_1 - s_{r^*}; Y^\varepsilon(s_{r^*}; x)) \in B_{(1/2)\varepsilon^{2\gamma}}(\phi^{\pm}).$$

Since in addition

$$\{ \sup_{t\in[s_{r*},T_1]} \|Y^{\varepsilon}(t;x) - u(t-s_{r*};Y^{\varepsilon}(s_{r*};x))\| \leqslant (1/2)\varepsilon^{2\gamma}\} \subseteq E_x,$$

we infer that

$$\begin{aligned} Y^{\varepsilon}(T_1,x) &= (Y^{\varepsilon}(T_1,x) - u(T_1-s_{r*},Y^{\varepsilon}(s_{r*};x))) + u(T_1-s_{r*},Y^{\varepsilon}(s_{r*};x)) \\ &\in B_{(1/2)\varepsilon^{2\gamma}}(0) + B_{(1/2)\varepsilon^{2\gamma}}(\phi^{\pm}) = B_{\varepsilon^{2\gamma}}(\phi^{\pm}). \end{aligned}$$

This proves (5.10).

2. On A_x, on which in particular $Y^{\varepsilon}(T_1;x)+\varepsilon W_1 \in \tilde{D}^{\pm}(\varepsilon^{\gamma})$ is valid, we infer with the help of paragraph (1) on $E_x \cap \{T_1 \geqslant s_{r*} + T_{rec} + \kappa\gamma|\ln\varepsilon|\}$, that

$$\varepsilon W_1 = Y^{\varepsilon}(T_1;x) + \varepsilon W_1 - Y^{\varepsilon}(T_1;x) \in \tilde{D}^{\pm}(\varepsilon^{\gamma}) - B_{\varepsilon^{2\gamma}}(\phi^{\pm}) \subseteq D_0^{\pm},$$

showing (i).

3. On B_x we have

$$Y^{\varepsilon}(T_1;x) + \varepsilon W_1 \in (\tilde{D}^{\pm}(\varepsilon^{\gamma}))^c.$$

Hence by paragraph (1), the event $B_x \cap E_x \cap \{T_1 \geqslant s_{r*} + T_{rec} + \kappa\gamma|\ln\varepsilon|\}$ implies

$$\varepsilon W_1 = Y^{\varepsilon}(T_1;x) + \varepsilon W_1 - Y^{\varepsilon}(T_1;x) \in (\tilde{D}_0^{\pm}(\varepsilon^{\gamma}))^c - B_{\varepsilon^{2\gamma}}(\phi^{\pm}) \subseteq (\hat{D}_0^{\pm}(\varepsilon^{\gamma}))^c,$$

proving (ii).

4. On C_x, we have

$$Y^{\varepsilon}(T_1;x) + \varepsilon W_1 \in \tilde{D}^{\pm}(\varepsilon^{\gamma}) \setminus \hat{D}^{\pm}(\varepsilon^{\gamma}).$$

Therefore on $C_x \cap E_x \cap \{T_1 \geqslant s_{r*} + T_{rec} + \kappa\gamma|\ln\varepsilon|\}$ the relationship

$$\varepsilon W_1 \in \left(\tilde{D}^{\pm}(\varepsilon^{\gamma}) \setminus \hat{D}^{\pm}(\varepsilon^{\gamma})\right) - B_{\varepsilon^{2\gamma}}(\phi^{\pm}) \subseteq D_0^{*}(\varepsilon^{\gamma})$$

follows, proving (iii).

5. On B_x, we know that $Y^{\varepsilon}(T_1;x)+\varepsilon W_1 \notin \tilde{D}^{\pm}(\varepsilon^{\gamma})$. But $T_1 \geqslant s_{r*} + T_{rec} + \kappa\gamma|\ln\varepsilon|$ and $\|\varepsilon W_1\| < (1/2)\varepsilon^{2\gamma}$ additionally entail that

$$Y^{\varepsilon}(T_1;x) + \varepsilon W_1 \in B_{(3/2)\varepsilon^{2\gamma}}(\phi^{\pm}) \cap \left(\tilde{D}^{\pm}(\varepsilon^{\gamma})\right)^c.$$

This set is empty for sufficiently small $\varepsilon > 0$, proving (iv).

6. On C_x, we have $Y^{\varepsilon}(T_1;x) + \varepsilon W_1 \in \tilde{D}^{\pm}(\varepsilon^{\gamma}) \setminus \hat{D}^{\pm}(\varepsilon^{\gamma})$. Imposing $T_1 \geqslant s_{r*} + T_{rec} + \kappa\gamma|\ln\varepsilon|$ and $\|\varepsilon W_1\| < (1/2)\varepsilon^{2\gamma}$ additionally leads to the intersection

$$\varepsilon W_1 \in \left(\tilde{D}^{\pm}(\varepsilon^{\gamma}) \setminus \hat{D}^{\pm}(\varepsilon^{\gamma})\right) \cap B_{(3/2)\varepsilon^{2\gamma}}(\phi^{\pm}) \subseteq (\hat{D}^{\pm}(\varepsilon^{\gamma}))^c \cap B_{(3/2)\varepsilon^{2\gamma}}(\phi^{\pm}),$$

which is empty for sufficiently small $\varepsilon > 0$. This proves (v).

7. First note that since $\hat{D}^{\pm}(\varepsilon^{\gamma}) \subseteq \tilde{D}^{\pm}(\varepsilon^{\gamma})$ for $\varepsilon > 0$ small enough paragraph (1) remains valid, i.e. $E_x \cap \{T_1 \geqslant s_{r^*} + T_{rec} + \kappa\gamma|\ln(\varepsilon)|\}$ implies $Y^{\varepsilon}(T_1; x) \in B_{\varepsilon^{2\gamma}}(\phi^{\pm})$. We show in addition that for $\varepsilon > 0$ sufficiently small and $x \in \hat{D}^{\pm}(\varepsilon^{\gamma})$ the event $E_x \cap \{T_1 - s_{r^*} \geqslant T_{rec} + \kappa\gamma|\ln \varepsilon|\}$ implies that

$$Y^{\varepsilon}(s; x) \in \tilde{D}^{\pm}(\varepsilon^{\gamma}) \quad \forall\, s \in [0, T_1]. \tag{5.12}$$

Definition 2.9 ensures for $x \in \hat{D}^{\pm}(\varepsilon^{\gamma})$, $\theta \in L^{\infty}(0, \infty; H)$ and $\sup_{t \geqslant 0} \|\theta(t)\| \leqslant \varepsilon^{2\gamma}$ that

$$v_{\theta}(t; x) + \theta(t) \in \tilde{D}^{\pm}(\varepsilon^{\gamma}) \text{ for all } t \geqslant 0.$$

Hence, the event $\{\sup_{t \in [0, T_1 \wedge s_{r^*}]} \|\varepsilon\xi^*(t)\| \leqslant \varepsilon^{2\gamma}\}$, which by (5.2) is a subset of E_x, implies

$$Y^{\varepsilon}(t; x) = v_{\varepsilon\xi^*}(t; x) + \varepsilon\xi^*(t) \in \tilde{D}^{\pm}(\varepsilon^{\gamma}) \text{ for all } t \in [0, T_1 \wedge s_{r^*}]. \tag{5.13}$$

Take $\theta = \varepsilon\xi^*$ on $[0, T_1 \wedge s_{r^*}]$ and $\theta = 0$ on $[T_1 \wedge s_{r^*}, \infty)$. Then $v_{\theta}(t; x) = Y^{\varepsilon}(t; x)$ on $[0, T_1 \wedge s_{r^*}]$ and $v_{\theta}(t; x) = u(t - T_1 \wedge s_{r^*}; Y^{\varepsilon}(T_1 \wedge s_{r^*})$ on $[T_1 \wedge s_{r^*}, \infty)$. Hence, by Definition 2.9,

$$B_{\varepsilon^{2\gamma}}(u(t; Y^{\varepsilon}(T_1 \wedge s_{r^*}; x))) \subseteq \tilde{D}^{\pm}(\varepsilon^{\gamma}) \text{ for } t \geqslant 0. \tag{5.14}$$

Since in addition to (5.14) by definition

$$E_x \subseteq \{ \sup_{t \in [s_{r^*}, T_1]} \|Y^{\varepsilon}(t; x) - u(t - s_{r^*}; Y^{\varepsilon}(s_{r^*}; x))\| \leqslant (1/2)\varepsilon^{2\gamma}\},$$

we infer for $t \in [s_{r^*}, T_1]$ that

$$\begin{aligned} Y^{\varepsilon}(t, x) &= (Y^{\varepsilon}(t, x) - u(t - s_{r^*}, Y^{\varepsilon}(s_{r^*}; x))) + u(t - s_{r^*}, Y^{\varepsilon}(s_{r^*}; x)) \\ &\in B_{(1/2)\varepsilon^{2\gamma}}(u(t - s_{r^*}, Y^{\varepsilon}(s_{r^*}; x))) \subseteq \tilde{D}^{\pm}(\varepsilon^{\gamma}). \end{aligned}$$

This proves (5.12). Note: Replacing in this reasoning $\tilde{D}^{\pm}(\varepsilon^{\pm})$ by $D^{\pm}(\varepsilon^{\pm})$ we obtain equally that for $\varepsilon > 0$ sufficiently small and $x \in \tilde{D}^{\pm}(\varepsilon^{\gamma})$ the event $E_x \cap \{T_1 - s_{r^*} \geqslant T_{rec} + \kappa\gamma|\ln \varepsilon|\}$ implies

$$Y^{\varepsilon}(s; x) \in D^{\pm}(\varepsilon^{\gamma}) \quad \forall\, s \in [0, T_1]. \tag{5.15}$$

8. By Parts 1 and 7 of the proof we know for $x \in \hat{D}^{\pm}(\varepsilon^{\gamma})$

$$E_x \cap \{T_1 \geqslant s_{r^*} + T_{rec} + \kappa\gamma |\ln \varepsilon|\}$$
$$\subseteq \{Y^\varepsilon(T_1; x) \in B_{\varepsilon^{2\gamma}}(\phi^\pm)\} \cap \{Y^\varepsilon(t; x) \in \tilde{D}^\pm(\varepsilon^\gamma), t \in [0, T_1]\}.$$

Hence on this set, $\varepsilon W_1 \in (D_0^\pm)^c$ implies additionally with the help of Lemma 2.11 that

$$Y^\varepsilon(T_1; x) + \varepsilon W_1 \in B_{\varepsilon^{2\gamma}}(\phi^\pm) + (D_0^\pm)^c \subseteq (\tilde{D}^\pm(\varepsilon^\gamma))^c,$$

which was missing to complete the event B_x. This proves (*vi*).

9. Analogously, by Parts 1 and 7 of the proof

$$E_x \cap \{T_1 \geqslant s_{r^*} + T_{rec} + \kappa\gamma |\ln \varepsilon|\}$$
$$\subseteq \{Y^\varepsilon(T_1; x) \in B_{\varepsilon^{2\gamma}}(\phi^\pm)\} \cap \{Y^\varepsilon(t; x) \in \tilde{D}^\pm(\varepsilon^\gamma), t \in [0, T_1]\}.$$

Thus the condition $\varepsilon W_1 \in D_0^\pm(\varepsilon^\gamma, \varepsilon^{2\gamma}, \varepsilon^{2\gamma}, \varepsilon^{2\gamma})$ yields with the help of Lemma 2.11

$$Y^\varepsilon(T_1; x) + \varepsilon W_1 \in B_{\varepsilon^{2\gamma}}(\phi^\pm) + D_0^\pm(\varepsilon^\gamma, \varepsilon^{2\gamma}, \varepsilon^{2\gamma}, \varepsilon^{2\gamma})$$
$$= B_{\varepsilon^{2\gamma}}(0) + \hat{D}^\pm(\varepsilon^\gamma, \varepsilon^{2\gamma}, \varepsilon^{2\gamma}, \varepsilon^{2\gamma}) \subseteq \hat{D}^\pm(\varepsilon^\gamma).$$

This proves (*vii*). □

The estimates presented in the preceding lemma can now be readily combined to provide full estimates of the events in terms of the first large jump time T_1, the large jump height W_1 and the perturbation event E_y^c on which deviations obtained from the small jump part are big.

Lemma 5.2 (Full estimates of the major events). *Let $T_{rec}, \kappa > 0$ given by Proposition 2.12 and $r^*, s_{r^*} > 0$ by Proposition 4.1. We denote the shift by time t on the path space for our Markov process X^ε resp. Y^ε by $\theta_t, t \geqslant 0$. Then for $\rho \in \left(\frac{1}{2}, 1\right)$, $\gamma \in (0, 1 - \rho)$ there exists $\varepsilon_0 > 0$ such that the following inequalities hold true for all $0 < \varepsilon \leqslant \varepsilon_0, \kappa > 0$ and $x \in \tilde{D}^\pm(\varepsilon^\gamma)$*

$$ix)\ \mathbf{1}(A_x) \leqslant \mathbf{1}\{\varepsilon W_1 \in D_0^\pm\} + \mathbf{1}\{\|\varepsilon W_1\| \geqslant \frac{1}{2}\varepsilon^{2\gamma}\}\mathbf{1}\{T_1 < s_{r^*} + T_{rec} + \kappa\gamma |\ln \varepsilon|\} + \mathbf{1}(E_x^c),$$

$$x)\ \mathbf{1}(B_x) \leqslant \mathbf{1}\{\varepsilon W_1 \notin \hat{D}_0^\pm(\varepsilon^\gamma)\} + \mathbf{1}\{T_1 < s_{r^*} + T_{rec} + \kappa\gamma |\ln \varepsilon|\} + \mathbf{1}(E_x^c),$$

xi) $\sup_{y\in\hat{D}^{\pm}(\varepsilon^{\gamma})} \mathbf{1}\{Y^{\varepsilon}(s;y) \notin \tilde{D}^{\pm}(\varepsilon^{\gamma}) \textit{ for some } s \in (0,T_1)\}$

$$\leqslant \sup_{y\in\hat{D}^{\pm}(\varepsilon^{\gamma})} \mathbf{1}(E_y^c)+\mathbf{1}\{T_1 < s_{r*} + T_{rec} + \kappa\gamma|\ln\varepsilon|\},$$

xii) $\mathbf{1}(A_x)\mathbf{1}\{Y^{\varepsilon}(s;X^{\varepsilon}(0,x))\circ\theta_{T_1} \notin \tilde{D}^{\pm}(\varepsilon^{\gamma}) \textit{ for some } s \in (0,t_2)\}$

$$\leqslant \mathbf{1}\{\varepsilon W_1 \in D_0^*(\varepsilon^{\gamma})\} + \mathbf{1}\{T_1 < s_{r*} + T_{rec} + \kappa\gamma|\ln\varepsilon|\}$$

$$+ \sup_{y\in\tilde{D}^{\pm}(\varepsilon^{\gamma})} \mathbf{1}(E_y^c)\circ\theta_{T_1} + \mathbf{1}(E_x^c) + \mathbf{1}\{T_1 < s_{r*} + T_{rec} + \kappa\gamma|\ln\varepsilon|\}\circ\theta(T_1).$$

In the opposite sense it follows for $x \in \hat{D}^{\pm}(\varepsilon^{\gamma})$

xiii) $\mathbf{1}(A_x^-) \geqslant \mathbf{1}\{\varepsilon W_1 \in D_0^{\pm}(\varepsilon^{\gamma},\varepsilon^{2\gamma},\varepsilon^{2\gamma},\varepsilon^{2\gamma})\} - \mathbf{1}\{T_1 < s_{r*} + T_{rec} + \kappa\gamma|\ln\varepsilon|\}$
$- \mathbf{1}(E_x^c),$

xiv) $\mathbf{1}(B_x) \geqslant \mathbf{1}\{\varepsilon W_1 \notin \tilde{D}_0^{\pm}\}(1 - \mathbf{1}\{T_1 < s_{r*} + T_{rec} + \kappa\gamma|\ln\varepsilon|\}) - \mathbf{1}(E_x^c).$

In particular for $x \in \hat{D}^{\pm}(\varepsilon^{\gamma})$

xv) $\mathbf{1}(A_x^- \cap A^{\diamond}) \geqslant \mathbf{1}\{\varepsilon W_1 \in D_0^{\pm}(\varepsilon^{\gamma},\varepsilon^{2\gamma},\varepsilon^{2\gamma},\varepsilon^{2\gamma})\} - \mathbf{1}\{T_1 < s_{r*} + T_{rec} + \kappa\gamma|\ln\varepsilon|\}$
$- \mathbf{1}(E_x^c),$

xvi) $\mathbf{1}(B_x \cap B^{\diamond}) \geqslant \mathbf{1}\{\varepsilon W_1 \notin D_0^{\pm}\}(1 - \mathbf{1}\{T_1 < s_{r*} + T_{rec} + \kappa\gamma|\ln\varepsilon|\}) - \mathbf{1}(E_x^c).$

Proof. We drop the superscript $\pm$ for convenience.

1. After a repartition of the event A_x for $x \in \tilde{D}(\varepsilon^{\gamma})$ we exploit Lemma 5.1 (i) and the fact that $B_{1/2\varepsilon^{2\gamma}}(0) \subseteq D_0$ in the third step of

$$\begin{aligned}
\mathbf{1}(A_x) &\leqslant \mathbf{1}(A_x)\mathbf{1}(E_x) + \mathbf{1}(E_x^c)\\
&\leqslant \mathbf{1}(A_x)\mathbf{1}(E_x)\mathbf{1}\{\varepsilon\|W_1\| \geqslant (1/2)\varepsilon^{2\gamma}\}\mathbf{1}\{T_1 \geqslant s_{r*} + T_{rec} + \kappa\gamma|\ln\varepsilon|\}\\
&\quad + \mathbf{1}(A_x)\mathbf{1}(E_x)\mathbf{1}\{\varepsilon\|W_1\| \geqslant (1/2)\varepsilon^{2\gamma}\}\mathbf{1}\{T_1 < s_{r*} + T_{rec} + \kappa\gamma|\ln\varepsilon|\}\\
&\quad + \mathbf{1}\{\varepsilon\|W_1\| < (1/2)\varepsilon^{2\gamma}\} + \mathbf{1}(E_x^c)\\
&\leqslant \mathbf{1}\{\varepsilon\|W_1\| \geqslant (1/2)\varepsilon^{2\gamma}\}\mathbf{1}\{\varepsilon W_1 \in D_0\}\\
&\quad + \mathbf{1}\{\varepsilon\|W_1\| \geqslant (1/2)\varepsilon^{2\gamma}\}\mathbf{1}\{T_1 < s_{r*} + T_{rec} + \kappa\gamma|\ln\varepsilon|\}\\
&\quad + \mathbf{1}\{\varepsilon\|W_1\| < (1/2)\varepsilon^{2\gamma}\}\mathbf{1}\{\varepsilon W_1 \in D_0\} + \mathbf{1}(E_x^c)\\
&= \mathbf{1}\{\varepsilon W_1 \in D_0\}\\
&\quad + \mathbf{1}\{\varepsilon\|W_1\| \geqslant (1/2)\varepsilon^{2\gamma}\}\mathbf{1}\{T_1 < s_{r*} + T_{rec} + \kappa\gamma|\ln\varepsilon|\} + \mathbf{1}(E_x^c).
\end{aligned}$$

This proves (ix).

2. In the same way we decompose B_x for $x \in \tilde{D}(\varepsilon^\gamma)$ and use Lemma 5.1 (*ii*) and (*iv*) in the second estimate to get

$$\begin{aligned}
\mathbf{1}(B_x) &\leqslant \mathbf{1}(E_x^c) \\
&\quad + \mathbf{1}(B_x)\mathbf{1}(E_x)\mathbf{1}\{\|\varepsilon W_1\| \geqslant (1/2)\varepsilon^{2\gamma}\}\mathbf{1}\{T_1 \geqslant s_{r^*} + T_{rec} + \kappa\gamma|\ln\varepsilon|\} \\
&\quad + \mathbf{1}(B_x)\mathbf{1}(E_x)\mathbf{1}\{\|\varepsilon W_1\| \geqslant (1/2)\varepsilon^{2\gamma}\}\mathbf{1}\{T_1 < s_{r^*} + T_{rec} + \kappa\gamma|\ln\varepsilon|\} \\
&\quad + \mathbf{1}(B_x)\mathbf{1}(E_x)\mathbf{1}\{\|\varepsilon W_1\| < (1/2)\varepsilon^{2\gamma}\}\mathbf{1}\{T_1 \geqslant s_{r^*} + T_{rec} + \kappa\gamma|\ln\varepsilon|\} \\
&\quad + \mathbf{1}(B_x)\mathbf{1}(E_x)\mathbf{1}\{\|\varepsilon W_1\| < (1/2)\varepsilon^{2\gamma}\}\mathbf{1}\{T_1 < s_{r^*} + T_{rec} + \kappa\gamma|\ln\varepsilon|\} \\
&\leqslant \mathbf{1}(E_x^c) + \mathbf{1}\{\varepsilon W_1 \notin \hat{D}_0(\varepsilon^\gamma)\} \\
&\quad + \mathbf{1}\{\|\varepsilon W_1\| \geqslant (1/2)\varepsilon^{2\gamma}\}\mathbf{1}\{T_1 < s_{r^*} + T_{rec} + \kappa\gamma|\ln\varepsilon|\} \\
&\quad + 0 + \mathbf{1}\{\|\varepsilon W_1\| < (1/2)\varepsilon^{2\gamma}\}\mathbf{1}\{T_1 < s_{r^*} + T_{rec} + \kappa\gamma|\ln\varepsilon|\} \\
&= \mathbf{1}(E_x^c) + \mathbf{1}\{\varepsilon W_1 \in (\tilde{D}_0(\varepsilon^\gamma)^c\} + \mathbf{1}\{T_1 < s_{r^*} + T_{rec} + \kappa\gamma|\ln\varepsilon|\}.
\end{aligned}$$

Hence (x) is shown.

3. Part 7 of Lemma 5.1 establishes that for $y \in \hat{D}(\varepsilon^\gamma) \subseteq \tilde{D}(\varepsilon^\gamma)$ the intersection $E_y \cap \{T_1 \geqslant s_{r^*} + T_{rec} + \kappa\gamma|\ln\varepsilon|\}$ implies

$$Y^\varepsilon(t; y) \in \tilde{D}(\varepsilon^\gamma) \quad \text{for } t \in [0, T_1].$$

Hence, by contraposition

$$\mathbf{1}\{Y^\varepsilon(s; y) \notin \tilde{D}(\varepsilon^\gamma) \text{ for some } s \in (0, T_1)\} \leqslant \mathbf{1}(E_y^c) + \mathbf{1}\{T_1 < s_{r^*} + T_{rec} + \kappa\gamma|\ln\varepsilon|\}.$$

Passing to the supremum on the right and left hand side proves (xi).

4. In this tedious estimate we have to take into account the evolution of the solution trajectory over two adjacent big jump intervals. Part 7 of the proof of Lemma 5.1 guarantees that for $x \in \tilde{D}^{\pm}(\varepsilon^\gamma)$

$$\begin{aligned}
&\mathbf{1}(A_x)\mathbf{1}\{Y^{\varepsilon,2}(s; X^\varepsilon(T_1, x)) \notin \tilde{D}(\varepsilon^\gamma) \text{ for some } s \in (0, t_2)\} \\
&\quad = \mathbf{1}\{Y^\varepsilon(s; x) \in \tilde{D}(\varepsilon^\gamma) \text{ for } s \in [0, t_1] \text{ and } X^\varepsilon(T_1; x) \in \hat{D}(\varepsilon^\gamma)\} \\
&\qquad \cdot \mathbf{1}\{Y^{\varepsilon,2}(s; X^\varepsilon(T_1; x)) \notin \tilde{D}(\varepsilon^\gamma) \text{ for some } s \in (0, t_2)\} \\
&\qquad + \mathbf{1}\{Y^\varepsilon(s; x) \in \tilde{D}(\varepsilon^\gamma) \text{ for } s \in [0, t_1] \text{ and } X^\varepsilon(T_1; x) \in \left(\tilde{D}(\varepsilon^\gamma) \setminus \hat{D}(\varepsilon^\gamma)\right)\} \\
&\qquad \cdot \mathbf{1}\{Y^{\varepsilon,2}(s; X^\varepsilon(T_1; x)) \notin \tilde{D}(\varepsilon^\gamma) \text{ for some } s \in [0, t_2]\} \\
&\quad \leqslant \sup_{z \in \hat{D}(\varepsilon^\gamma)} \mathbf{1}\{Y^{\varepsilon,2}(s; z) \notin \tilde{D}(\varepsilon^\gamma) \text{ for some } s \in [0, t_2]\} \\
&\qquad + \mathbf{1}\{Y^\varepsilon(s; x) \in \tilde{D}(\varepsilon^\gamma) \text{ for } s \in [0, t_1] \text{ and } X^\varepsilon(T_1; x) \in \left(\tilde{D}(\varepsilon^\gamma) \setminus \hat{D}(\varepsilon^\gamma)\right)\} \\
&\quad = \sup_{z \in \hat{D}(\varepsilon^\gamma)} \mathbf{1}\{Y^\varepsilon(s; z) \notin \tilde{D}(\varepsilon^\gamma) \text{ for some } s \in (0, T_1)\} \circ \theta_{T_1} + \mathbf{1}(C_x).
\end{aligned} \tag{5.16}$$

Now we repeat the arguments employed for Part 2, replacing $\tilde{D}_0^c(\varepsilon^\gamma)$ by $D_0^*(\varepsilon^\gamma)$ defined before. We may exploit Lemma 5.1 (*iii*) here and obtain

$$\begin{aligned}
\mathbf{1}(C_x) &\leqslant \mathbf{1}(E_x^c) \\
&\quad + \mathbf{1}(C_x)\mathbf{1}(E_x)\mathbf{1}\{\|\varepsilon W_1\| \geqslant (1/2)\varepsilon^{2\gamma}\}\mathbf{1}\{T_1 \geqslant s_{r^*} + T_{rec} + \kappa\gamma|\ln\varepsilon|\} \\
&\quad + \mathbf{1}(C_x)\mathbf{1}(E_x)\mathbf{1}\{\|\varepsilon W_1\| \geqslant (1/2)\varepsilon^{2\gamma}\}\mathbf{1}\{T_1 < s_{r^*} + T_{rec} + \kappa\gamma|\ln\varepsilon|\} \\
&\quad + \mathbf{1}(C_x)\mathbf{1}(E_x)\mathbf{1}\{\|\varepsilon W_1\| < (1/2)\varepsilon^{2\gamma}\}\mathbf{1}\{T_1 \geqslant s_{r^*} + T_{rec} + \kappa\gamma|\ln\varepsilon|\} \\
&\quad + \mathbf{1}(C_x)\mathbf{1}(E_x)\mathbf{1}\{\|\varepsilon W_1\| < (1/2)\varepsilon^{2\gamma}\}\mathbf{1}\{T_1 < s_{r^*} + T_{rec} + \kappa\gamma|\ln\varepsilon|\} \\
&\leqslant \mathbf{1}(E_x^c) + \mathbf{1}\left\{\varepsilon W_1 \in \left(\tilde{D}_0(\varepsilon^\gamma) \setminus \hat{D}_0(\varepsilon^\gamma)\right) + B_{\varepsilon^{2\gamma}}(0)\right\} \\
&\quad + \mathbf{1}\{\|\varepsilon W_1\| \geqslant (1/2)\varepsilon^{2\gamma}\}\mathbf{1}\{T_1 < s_{r^*} + T_{rec} + \kappa\gamma|\ln\varepsilon|\} + 0 \\
&\quad + \mathbf{1}\{\|\varepsilon W_1\| < (1/2)\varepsilon^{2\gamma}\}\mathbf{1}\{T_1 < s_{r^*} + T_{rec} + \kappa\gamma|\ln\varepsilon|\} \\
&\leqslant \mathbf{1}(E_x^c) + \mathbf{1}\{\varepsilon W_1 \in D_0^*(\varepsilon^\gamma)\} + \mathbf{1}\{T_1 < s_{r^*} + T_{rec} + \kappa\gamma|\ln\varepsilon|\}.
\end{aligned} \tag{5.17}$$

Hence collecting the estimates (5.16) and (5.17) and applying (xi) we obtain

$$\begin{aligned}
&\mathbf{1}(A_x)\mathbf{1}\{Y^\varepsilon(s; X^\varepsilon(0,x)) \circ \theta_{T_1} \notin \tilde{D}(\varepsilon^\gamma) \text{ for some } s \in [0, T_1]\} \\
&\quad \leqslant \mathbf{1}\{\varepsilon W_1 \in D_0^*(\varepsilon^\gamma)\} + \mathbf{1}\{T_1 < s_{r^*} + T_{rec} + \kappa\gamma|\ln\varepsilon|\} \\
&\qquad + \sup_{z\in\hat{D}(\varepsilon^\gamma)} \mathbf{1}\{Y^\varepsilon(s; z) \circ \theta_{T^1} \notin \tilde{D}(\varepsilon^\gamma) \text{ for some } s \in [0, T_1]\} + \mathbf{1}(E_x^c) \\
&\quad \leqslant \mathbf{1}\{\varepsilon W_1 \in D_0^*(\varepsilon^\gamma)\} + \mathbf{1}\{T_1 < s_{r^*} + T_{rec} + \kappa\gamma|\ln\varepsilon|\} \\
&\qquad + \sup_{y\in\hat{D}(\varepsilon^\gamma)} \mathbf{1}(E_y^c) \circ \theta_{T_1} + \mathbf{1}(E_x^c) + \mathbf{1}\{T_1 < s_{r^*} + T_{rec} + \kappa\gamma|\ln\varepsilon|\} \circ \theta(T_1).
\end{aligned}$$

This proves (*xii*).

5. For $x \in \hat{D}(\varepsilon^\gamma)$ Lemma 5.1 (*vii*) states

$$\mathbf{1}(A_x^-) \geqslant \mathbf{1}(E_x)\mathbf{1}\{T_1 \geqslant s_{r^*} + T_{rec} + \kappa\gamma|\ln\varepsilon|\}\mathbf{1}\{\varepsilon W_1 \in D_0^\pm(\varepsilon^\gamma, \varepsilon^{2\gamma}, \varepsilon^{2\gamma}, \varepsilon^{2\gamma})\}.$$

With the help of the elementary inequality

$$\mathbf{1}(C_1)\mathbf{1}(C_2) = \mathbf{1}(C_1)(1 - \mathbf{1}(C_2^c)) \geqslant \mathbf{1}(C_1) - \mathbf{1}(C_2^c)$$

valid for arbitrary sets C_1, C_2 we obtain

$$\mathbf{1}(A_x^-) \geqslant \mathbf{1}\{\varepsilon W_1 \in D_0^\pm(\varepsilon^\gamma, \varepsilon^{2\gamma}, \varepsilon^{2\gamma}, \varepsilon^{2\gamma})\} - \mathbf{1}(E_x^c) - \mathbf{1}\{T_1 < s_{r^*} + T_{rec} + \kappa\gamma|\ln\varepsilon|\}.$$

This proves statement (*xiii*).

6. To obtain the last estimate (*xiv*) for $x \in \hat{D}(\varepsilon^\gamma)$, we use Lemma 5.1 (*vi*) which yields

$$\begin{aligned}\mathbf{1}(B_x) &\geqslant \mathbf{1}(E_x)\mathbf{1}\{T_1 \geqslant s_{r^*} + T_{rec} + \kappa\gamma|\ln\varepsilon|\}\mathbf{1}\{\varepsilon W_1 \notin D_0\} \\ &\geqslant \mathbf{1}\{\varepsilon W_1 \notin D_0\}(1 - \mathbf{1}\{T_1 < s_{r^*} + T_{rec} + \kappa\gamma|\ln\varepsilon|\}) - \mathbf{1}(E_x^c). \quad (5.18)\end{aligned}$$

7. Since by Lemma 5.1 (*vii*)

$$\mathbf{1}(A_x^-) \geqslant \mathbf{1}(E_x)\mathbf{1}\{T_1 \geqslant s_{r^*} + T_{rec} + \kappa\gamma|\ln\varepsilon|\}\mathbf{1}\{\varepsilon W_1 \in D_0^{\pm}(\varepsilon^{\gamma}, \varepsilon^{2\gamma}, \varepsilon^{2\gamma}, \varepsilon^{2\gamma})\}$$

and $\{\varepsilon W_1 \in D_0^{\pm}(\varepsilon^{\gamma}, \varepsilon^{2\gamma}, \varepsilon^{2\gamma}, \varepsilon^{2\gamma})\} \subseteq \{\varepsilon W_1 \in D_0^{\pm}\} = A_1^{\diamond}$, we obtain

$$\mathbf{1}(A_x^- \cap A_1^{\diamond}) \geqslant \mathbf{1}(E_x)\mathbf{1}\{T_1 \geqslant s_{r^*} + T_{rec} + \kappa\gamma|\ln\varepsilon|\}\mathbf{1}\{\varepsilon W_1 \in D_0^{\pm}(\varepsilon^{\gamma}, \varepsilon^{2\gamma}, \varepsilon^{2\gamma}, \varepsilon^{2\gamma})\} \quad (5.19)$$

and by the same reasoning as for part 5

$$\begin{aligned}&\mathbf{1}(A_x^- \cap A_1^{\diamond}) \\ &\geqslant \mathbf{1}\{\varepsilon W_1 \in D_0^{\pm}(\varepsilon^{\gamma}, \varepsilon^{2\gamma}, \varepsilon^{2\gamma}, \varepsilon^{2\gamma})\} - \mathbf{1}\{T_1 < s_{r^*} + T_{rec} + \kappa\gamma|\ln\varepsilon|\} - \mathbf{1}(E_x^c). \quad (5.20)\end{aligned}$$

This shows inequality (*xv*).

8. Similarly since by Lemma 5.1 (*vi*)

$$\mathbf{1}(B_x) \geqslant \mathbf{1}(E_x)\mathbf{1}\{T_1 \geqslant s_{r^*} + T_{rec} + \kappa\gamma|\ln\varepsilon|\}\mathbf{1}(B_1^{\diamond}),$$

it follows

$$\mathbf{1}(B_x \cap B_1^{\diamond}) \geqslant \mathbf{1}(E_x)\mathbf{1}\{T_1 \geqslant s_{r^*} + T_{rec} + \kappa\gamma|\ln\varepsilon|\}\mathbf{1}(B_1^{\diamond}), \quad (5.21)$$

giving the desired estimate (*xvi*)

$$\mathbf{1}(B_x \cap B_1^{\diamond}) \geqslant \mathbf{1}\{\varepsilon W_1 \notin D_0^{\pm}\}(1 - \mathbf{1}\{T_1 < s_{r^*} + T_{rec} + \kappa\gamma|\ln\varepsilon|\}) - \mathbf{1}(E_x^c). \quad (5.22)$$

□

5.1.2 Hypotheses on the Lévy Measure

In this subsection we present two hypothesis on the Lévy measure with respect to the (infinite-dimensional) geometry of the domains of attraction of the deterministic dynamical system and examine their consequences, in particular of (H.2), which will be used explicitly in the proofs of the next section.

We use the following hypotheses

Let μ be the limiting measure of the symmetric, regularly varying Lévy measure ν of index $\alpha \in (0, 2)$, for details see Sect. 3.6. We assume the following hypotheses.

(H.1) Non-trivial transitions:

$$\mu\left(\left(D_0^{\pm}\right)^c\right) > 0.$$

This condition excludes that the system remains in one domain of attraction.

(H.2) Non-degenerate limiting measure:

$$\mu\left(\mathscr{S} - \phi^{\pm}\right) = 0. \tag{5.23}$$

This condition ensures that for the process sitting exactly on one of the stable states $\phi^{\pm}$ there is no probability mass for a "large" jump W_i exactly onto the separatrix $\mathscr{S} = H \setminus (D^+ \cup D^-)$, where it might get stuck.

The following corollary shows that due to the set monotonicity $\varepsilon \mapsto D_0^{\pm}(\varepsilon^{\gamma}, \varepsilon^{2\gamma}, \varepsilon^{2\gamma}, \varepsilon^{2\gamma})$ and the respective continuity of the limiting measure μ for the Lévy measure ν this is the statement of the previous phrase is robust under small (ε-dependent) perturbations.

Lemma 5.3. *Fix $\gamma \in (0, 1)$. Then for each $\eta > 0$ there is $\varepsilon_0 > 0$ such that for $\varepsilon \in (0, \varepsilon_0)$ and $k \in \{+, -\}$ follows*

$$\mu\left(H \setminus \left((D^+(\varepsilon^{\gamma}, \varepsilon^{2\gamma}, \varepsilon^{2\gamma}, \varepsilon^{2\gamma}) \cup D^-(\varepsilon^{\gamma}, \varepsilon^{2\gamma}, \varepsilon^{2\gamma}, \varepsilon^{2\gamma})) + B_{\varepsilon^{2\gamma}}(0)\right) - \phi^k\right) < \eta. \tag{5.24}$$

Proof. By Lemma 2.10 we know that

$$D^{\pm} = \bigcup_{\varepsilon > 0} D^{\pm}(\varepsilon^{\gamma}, \varepsilon^{2\gamma}, \varepsilon^{2\gamma}, \varepsilon^{2\gamma})$$

and by construction $\varepsilon \mapsto D^{\pm}(\varepsilon^{\gamma}, \varepsilon^{2\gamma}, \varepsilon^{2\gamma}, \varepsilon^{2\gamma})$ is growing mononically growing with respect to set inclusions. Hence

$$\varepsilon \mapsto H \setminus \left((D^+(\varepsilon^{\gamma}, \varepsilon^{2\gamma}, \varepsilon^{2\gamma}, \varepsilon^{2\gamma}) \cup D^-(\varepsilon^{\gamma}, \varepsilon^{2\gamma}, \varepsilon^{2\gamma}, \varepsilon^{2\gamma})) + B_{\varepsilon^{2\gamma}}(0)\right) - \phi^k$$

is mononically decreasing and with the help of the monotonicity of the set function μ and Hypothesis (H.2)

$$\begin{aligned}
&\lim_{\varepsilon \to 0+} \mu\left(H \setminus \left((D^+(\varepsilon^{\gamma}, \varepsilon^{2\gamma}, \varepsilon^{2\gamma}, \varepsilon^{2\gamma}) \cup D^-(\varepsilon^{\gamma}, \varepsilon^{2\gamma}, \varepsilon^{2\gamma}, \varepsilon^{2\gamma})) + B_{\varepsilon^{2\gamma}}(0)\right) - \phi^k\right) \\
&\quad = \mu((H \setminus (D^+ \cup D^-)) - \phi^k) \\
&\quad = \mu(\mathscr{S} - \phi^k) = 0.
\end{aligned}$$

□

Hypothesis **(H.2)** and Lemma 5.3 imply a sequence of more sophisticated estimates of similar type, which will be applied directly in the sequel.

Lemma 5.4. *Assume that Hypothesis (H.2) is true and* $\gamma \in (0,1)$. *Then for any* $\eta > 0$ *we can choose* $\varepsilon_0 > 0$ *small enough such that for all* $0 < \varepsilon \leqslant \varepsilon_0$

$$i)\ \mu\left(\left(\tilde{D}_0^{\pm}(\varepsilon^{\gamma})\right)^c \setminus \left(D_0^{\pm}\right)^c\right) < \eta,$$

$$ii)\ \mu\left(\left(D_0^{\pm}(\varepsilon^{\gamma}) \setminus \tilde{D}_0^{\pm}(\varepsilon^{\gamma})\right) + B_{\varepsilon^{2\gamma}}(0)\right) < \eta,$$

$$iii)\ \mu\left(D_0^{\pm} \setminus D_0^{\pm}(\varepsilon^{\gamma}, \varepsilon^{2\gamma}, \varepsilon^{2\gamma}, \varepsilon^{2\gamma})\right) < \eta,$$

$$iv)\ \mu\left((D^{\pm})^c \setminus \tilde{D}^{\mp}(\varepsilon^{\gamma}) - \phi^{\pm}\right) < \eta,$$

$$v)\ \mu\left(\left(\hat{D}_0^{\pm}(\varepsilon^{\gamma})\right)^c \setminus \left(D_0^{\pm}\right)^c\right) < \eta.$$

The claims follow directly from the set inclusions in Sect. 2.2.1, Lemma 5.3 and again the set monotonicity of $\varepsilon \mapsto D^{\pm}(\varepsilon^{\gamma})$, $\tilde{D}^{\pm}(\varepsilon^{\gamma})$, etc.

Equipped with estimates of the major events by analytically accessible handy ones containing only information about the time and height of the first big jump and the deviations of the small jump part from the deterministic solution before the first big jump time, we can study their asymptotic behavior. It will turn out that only the large jump event stipulating W_1 to leave $D_0^{\pm}$ or its reduced versions will be asymptotically relevant. This is rigorously stated in the following lemma.

Lemma 5.5 (Asymptotic behavior of large jump events).
Assume that Hypotheses (H.1) and (H.2) are satisfied and let $1/2 < \rho < 1 - 2\gamma$ *fixed. Then for any* $a > 0$ *there is* $\varepsilon_0 = \varepsilon_0(a) > 0$ *such that for all* $0 < \varepsilon \leqslant \varepsilon_0$

$$I)\ \frac{\mu\left((D_0^{\pm})^c\right)}{\mu(B_1^c(0))}\varepsilon^{\alpha(1-\rho)+a} \leqslant \frac{\lambda^{\pm}(\varepsilon)}{\beta_\varepsilon} \leqslant \frac{\mu((D_0^{\pm})^c)}{\mu(B_1^c(0))}\varepsilon^{\alpha(1-\rho)-a},$$

$$II)\ \mathbb{P}\left(\|\varepsilon W_1\| \geqslant (1/2)\varepsilon^{2\gamma}\right) \leqslant \varepsilon^{\alpha(1-\rho-2\gamma)-a},$$

Moreover, for any $C > 0$ *there is* $\varepsilon_0 = \varepsilon_0(C) > 0$ *such that for all* $0 < \varepsilon \leqslant \varepsilon_0$

$$III)\quad \mathbb{P}\left(\varepsilon W_1 \in (\tilde{D}_0^{\pm}(\varepsilon^{\gamma}))^c\right) \leqslant (1+C)\frac{\lambda^{\pm}(\varepsilon)}{\beta_\varepsilon},$$

$$IV)\quad \mathbb{P}\left(\varepsilon W_1 \in D_0^{*}(\varepsilon^{\gamma})\right) \leqslant C\frac{\lambda^{\pm}(\varepsilon)}{\beta_\varepsilon},$$

$$V)\quad \mathbb{P}(\varepsilon W_1 \in (D_0^{\pm}(\varepsilon^{\gamma}, \varepsilon^{2\gamma}, \varepsilon^{2\gamma}, \varepsilon^{2\gamma}))^c) \leqslant (1+C)\frac{\lambda^{\pm}(\varepsilon)}{\beta_\varepsilon},$$

$$VI)\quad \mathbb{P}\left(\varepsilon W_1 \in (\hat{D}_0^{\pm}(\varepsilon^{\gamma}))^c\right) \leqslant (1+C)\,\frac{\lambda^{\pm}(\varepsilon)}{\beta_\varepsilon}.$$

Proof. 1. For convenience we drop the exponent $\pm$.
By (3.30), we have

$$\frac{\lambda(\varepsilon)}{\beta_\varepsilon}\Big[\varepsilon^{\alpha(1-\rho)}\frac{\ell(\varepsilon^{-1})}{\ell(\varepsilon^{-\rho})}\frac{\mu(D_0^c)}{\mu(B_1^c(0))}\Big]^{-1} \to 1, \qquad \text{as } \varepsilon \to 0+.$$

(I) follows easily from Proposition 3.43.

2. By the choice $1/2 < \rho < 1-2\gamma$ for $0<\varepsilon\leqslant\varepsilon_0$ small enough

$$\mathbb{P}\left(\varepsilon\|W_1\| \geqslant (1/2)\varepsilon^{2\gamma}\right) = \frac{\nu\left(\left((1/2)\varepsilon^{2\gamma-1}\wedge\varepsilon^{-\rho}\right)B_1^c(0)\right)}{\nu\left((\varepsilon^{-\rho})\,B_1^c(0)\right)} = \frac{\nu\left((1/2)\varepsilon^{2\gamma-1}B_1^c(0)\right)}{\nu\left((\varepsilon^{-\rho})\,B_1^c(0)\right)}.$$

Thus

$$\mathbb{P}\left(\varepsilon\|W_1\| \geqslant (1/2)\varepsilon^{2\gamma}\right)\Big[(\frac{1}{2})^{-\alpha}\varepsilon^{\alpha(1-2\gamma-\rho)}\frac{\ell(\frac{1}{2}\varepsilon^{2\gamma-1})}{\ell(\varepsilon^{-\rho})}\Big] \to 1, \qquad \text{as } \varepsilon \to 0+.$$

We use again Proposition 3.43 to conclude that *(II)* holds.

3. We write

$$\mathbb{P}\left(\varepsilon W_1 \in \tilde{D}_0^c(\varepsilon^{\gamma})\right) - \frac{\lambda(\varepsilon)}{\beta_\varepsilon} = \frac{\nu\left(\frac{1}{\varepsilon}\tilde{D}_0^c(\varepsilon^{\gamma})\right) - \nu\left(\frac{1}{\varepsilon}D_0^c\right)}{\beta_\varepsilon} = \frac{\nu\left(\frac{1}{\varepsilon}\left(\tilde{D}_0^c(\varepsilon^{\gamma})\setminus D_0^c\right)\right)}{\beta_\varepsilon}.$$

Thus

$$\left(\mathbb{P}\left(\varepsilon W_1 \in \tilde{D}_0^c(\varepsilon^{\gamma})\right) - \frac{\lambda(\varepsilon)}{\beta_\varepsilon}\right)\Big[\varepsilon^{\alpha(1-\rho)}\frac{\ell(\varepsilon^{-1})}{\ell(\varepsilon^{-\rho})}\frac{\mu\left(\tilde{D}_0^c(\varepsilon^{\gamma})\setminus D_0^c\right)}{\mu(B_1^c(0))}\Big]^{-1}$$
$$\to 1, \qquad \text{as } \varepsilon \to 0+,$$

and hence

$$\left(\mathbb{P}\left(\varepsilon W_1 \in \tilde{D}_0^c(\varepsilon^{\gamma})\right) - \frac{\lambda(\varepsilon)}{\beta_\varepsilon}\right)\Big[\frac{\lambda(\varepsilon)}{\beta_\varepsilon}\frac{\mu\left(\tilde{D}_0^c(\varepsilon^{\gamma})\setminus D_0^c\right)}{\mu(D_0^c)}\Big]^{-1} \to 1, \qquad \text{as } \varepsilon \to 0+.$$

By Lemma 5.4 (i), for any $C>0$ there is $\varepsilon_0>0$ such that $\frac{\mu(\tilde{D}_0^c(\varepsilon^{\gamma})\setminus D_0^c)}{\mu(D_0^c)} < C$ for $0<\varepsilon\leqslant\varepsilon_0$. We conclude that *(III)* is true.

4. Analogously, using (H.2) and Lemma 5.4 (*ii*), we find that for $C > 0$ there exists $\varepsilon_0 > 0$ such that $\dfrac{\mu\left(D_0^*(\varepsilon^\gamma)\right)}{\mu(D_0^c)} < C$ for $\varepsilon \leqslant \varepsilon_0$. Due to the regular variation of ν we can write

$$\mathbb{P}\left(\varepsilon W_1 \in D_0^*(\varepsilon^\gamma)\right)\left[\varepsilon^{\alpha(1-\rho)}\frac{\ell(\varepsilon^{-1})}{\ell(\varepsilon^{-\rho})}\frac{\mu\left(\tilde{D}_0^*(\varepsilon^\gamma)\right)}{\mu(B_1^c(0))}\right]^{-1} \to 1, \qquad \text{as } \varepsilon \to 0+,$$

and therefore

$$\mathbb{P}\left(\varepsilon W_1 \in D_0^*(\varepsilon^\gamma)\right)\left[\frac{\lambda(\varepsilon)}{\beta_\varepsilon}\frac{\mu\left(\tilde{D}_0^*(\varepsilon^\gamma)\right)}{\mu(D_0^c)}\right]^{-1} \to 1, \qquad \text{as } \varepsilon \to 0+.$$

There exists $\varepsilon_0 > 0$ such that for $\varepsilon \leqslant \varepsilon_0$

$$\mathbb{P}\left(\varepsilon W_1 \in D_0^*(\varepsilon^\gamma)\right)\left[\frac{\lambda(\varepsilon)}{\beta_\varepsilon}\right]^{-1} < C.$$

Hence (*IV*) is proved.

5. The argument for (*V*) [resp. (*VI*)] is identical to the one for (*III*). We just have to replace $\tilde{D}_0^c(\varepsilon^\gamma) \setminus D_0^c$ by $D_0 \setminus D_0(\varepsilon^\gamma, \varepsilon^{2\gamma}, \varepsilon^{2\gamma}, \varepsilon^{2\gamma})$ (resp. $\hat{D}_0^c(\varepsilon^\gamma) \setminus D_0^c$) and use Lemma 5.4 (*iii*) [resp. (*v*)]. This yields the desired estimate for ε_0 small enough. □

5.2 Asymptotic Exit Times from Reduced Domains of Attraction

In this section we shall state and prove our main result about the asymptotic behavior of the exit time from the reduced domains of attraction of the equilibria of the Chafee–Infante equation. It will essentially describe the asymptotic behavior of the exit time's Laplace transform. Let us start with a remark concerning the constants appearing in the small deviations estimates in Chap. 4.

Definition 5.6. For $\gamma \in (0,1)$, $\varepsilon > 0$ and $X^\varepsilon(\cdot; x)$ the càdlàg mild solution of (1.2), with initial position $x \in \tilde{D}^\pm(\varepsilon^\gamma)$ we define the *first exit time from the reduced domain of attraction*

$$\tau_x^\pm(\varepsilon) := \inf\{t > 0 \mid X^\varepsilon(t; x) \notin \tilde{D}^\pm(\varepsilon^\gamma)\}.$$

The following fixing of constants links the small noise estimates in Chap. 4 with the estimates in Sect. 5.1, in particular with the control of the asymptotics of the probability of the event E_x^c (Fig. 5.1).

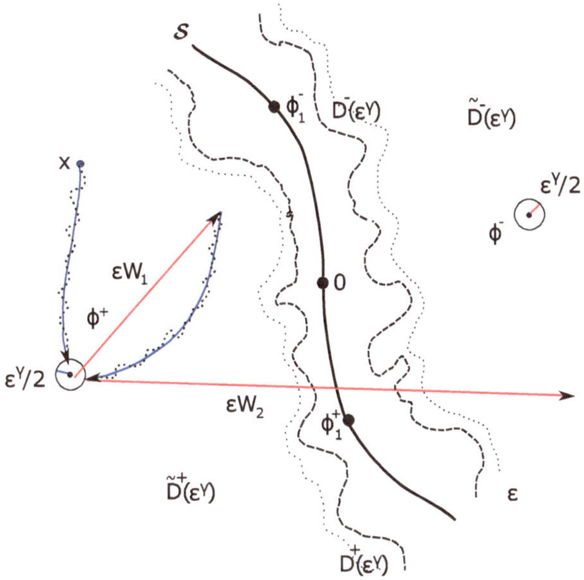

Fig. 5.1 Sketch of a typical first exit event from a reduced domain of attraction

Convention (C)

Let the Chafee–Infante parameter $\pi^2 < \lambda \neq (k\pi)^2$ *for* $k \in \mathbb{N}$ *be given, the constants* $T_{rec}, \kappa > 0$ *chosen in Proposition 2.12 and* $r^*, s_{r^*} > 0$ *be given by Proposition 4.1. For* $\alpha \in (0,2)$ *and* $\Gamma > 0$ *chosen large enough according to Lemma 4.8 in the proof of Proposition 4.5 we fix the constants* Θ, ρ, γ *satisfying*

$$0 < \Theta < \frac{2-\alpha}{2\alpha}, \quad \rho \in \left(\frac{1}{2}, \frac{2-\alpha}{2-(1-\Theta)\alpha}\right), \quad 0 < \gamma < \frac{(2-\alpha)(1-\rho)-\Theta\alpha\rho}{2(\Gamma+2)}. \tag{5.25}$$

Remark 5.7. • The details of the proof of Proposition 4.5 in Sect. 4.4 ensure the well-defined choice of the constants in Convention (C). In addition, Proposition 4.5 stays true for any constant $\tilde{\Gamma} > \Gamma$ and $\Theta < \frac{2-\alpha}{2\alpha}$.

• Convention (C) implies in particular that

$$2\gamma < \rho < 1 - 2\gamma \tag{5.26}$$

is satisfied, which was necessary in Lemma 5.5. In fact, the first point is evident since $\rho > 1/2 > 1/4 > \gamma$. For the second inequality we check easily

$$\rho + 2\gamma < \rho + \frac{(2-\alpha)(1-\rho)}{\Gamma+2} \leqslant \rho + 1 - \rho = 1.$$

The following theorem states that for all $\theta > -1$ the Laplace transform $\widehat{\lambda(\varepsilon)\tau_x^{\pm}(\varepsilon)}(\theta)$ of the normalized first exit time $\lambda(\varepsilon)\tau_x^{\pm}(\varepsilon)$ from the reduced domain of attraction $\tilde{D}^{\pm}(\varepsilon^{\gamma})$ of $D^{\pm}$ converges to $\frac{1}{1+\theta}$ as $\varepsilon \to 0+$. This establishes its convergence in law to an exponentially distributed random variable.

Theorem 5.8 (Asymptotic first exit time law). *Assume that Hypotheses (H.1) and (H.2) and Convention (C) are satisfied. Then for all $\theta > -1$ and $C \in (0, 1+\theta)$ there exists $\varepsilon_0 = \varepsilon_0(\theta) > 0$ such that for all $0 < \varepsilon \leqslant \varepsilon_0$*

$$\frac{1-C}{1+\theta+C} \leqslant \mathbb{E}\left[\inf_{x\in\hat{D}^{\pm}(\varepsilon^{\gamma})} \exp\left(-\theta\lambda^{\pm}(\varepsilon)\tau_x^{\pm}(\varepsilon)\right)\right]$$
$$\leqslant \mathbb{E}\left[\sup_{x\in\hat{D}^{\pm}(\varepsilon^{\gamma})} \exp\left(-\theta\lambda^{\pm}(\varepsilon)\tau_x^{\pm}(\varepsilon)\right)\right] \leqslant \frac{1+C}{1+\theta-C}.$$

The theorem is proved in Sects. 5.2.1 and 5.2.2. Its result implies a statement about the asymptotic behavior of the expected first exit time.

Corollary 5.9. *Under the assumptions of Theorem 5.8 we have*

$$\lim_{\varepsilon\to0+} \mathbb{E}\left[\inf_{x\in\hat{D}^{\pm}(\varepsilon^{\gamma})} \lambda^{\pm}(\varepsilon)\tau_x^{\pm}(\varepsilon)\right] = \lim_{\varepsilon\to0+} \mathbb{E}\left[\sup_{x\in\hat{D}^{\pm}(\varepsilon^{\gamma})} \lambda^{\pm}(\varepsilon)\tau_x^{\pm}(\varepsilon)\right] = 1. \tag{5.27}$$

Proof. By Theorem 5.8 which holds for $\theta > -1$, we know that $\lambda^{\pm}(\varepsilon)\tau_x(\varepsilon)$ converges in law to τ as $\varepsilon \to 0$, and τ has an exponential law with parameter 1. In addition, for $\theta < 0$

$$\mathbb{E}\left[e^{-\theta\left(\inf_{x\in\hat{D}^{\pm}(\varepsilon^{\gamma})}\lambda^{\pm}(\varepsilon)\tau_x^{\pm}(\varepsilon)\right)}\right]$$
$$\leqslant \mathbb{E}\left[e^{-\theta\left(\sup_{x\in\hat{D}^{\pm}(\varepsilon^{\gamma})}\lambda^{\pm}(\varepsilon)\tau_x^{\pm}(\varepsilon)\right)}\right] = \mathbb{E}\left[\sup_{x\in\hat{D}^{\pm}(\varepsilon^{\gamma})} e^{-\theta\lambda^{\pm}(\varepsilon)\tau_x^{\pm}(\varepsilon)}\right] \leqslant \frac{1+C}{1+\theta-C} < \infty$$

and hence $(\inf_{x\in\hat{D}^{\pm}(\varepsilon^{\gamma})}\lambda^{\pm}(\varepsilon)\tau_x^{\pm}(\varepsilon))_{0<\varepsilon\leqslant\varepsilon_0}$ and $(\sup_{x\in\hat{D}^{\pm}(\varepsilon^{\gamma})}\lambda^{\pm}(\varepsilon)\tau_x^{\pm}(\varepsilon))_{0<\varepsilon\leqslant\varepsilon_0}$ are uniformly integrable. For nonnegative random variables, convergence in law and uniform integrability implies convergence in expectation (for instance [Kal97], Lemma 4.11). This implies equality (5.27). □

In the sequel we shall construct a family of random variables, $(\bar{\tau}(\varepsilon))_{\varepsilon>0}$, such that in probability $\tau_x^{\pm}(\varepsilon)\lambda^{\pm}(\varepsilon) - \bar{\tau}(\varepsilon) \to 0$ for $\varepsilon \to 0+$.

Theorem 5.10 (Asymptotic first exit times in probability). *Assume that Hypotheses (H.1) and (H.2) and Convention (C) are satisfied. Then there is a family of random variables $(\bar{\tau}(\varepsilon))_{\varepsilon>0}$ with exponential law of parameter 1 (on the* *same probability space $(\Omega, \mathcal{F}, \mathbb{P})$ as the driving Lévy noise $(L(t))_{t\geqslant 0})$ such that in probability*

$$\lim_{\varepsilon\to 0+} \inf_{y\in \hat{D}^{\pm}(\varepsilon^{\gamma})} |\lambda^{\pm}(\varepsilon)\tau_y^{\pm} - \bar{\tau}(\varepsilon)| = \lim_{\varepsilon\to 0+} \inf_{y\in \hat{D}^{\pm}(\varepsilon^{\gamma})} |\lambda^{\pm}(\varepsilon)\tau_y^{\pm} - \bar{\tau}(\varepsilon)| = 0.$$

This theorem is proved in Sect. 5.2.3. Combining the preceding two theorems, we obtain the following main result on the first exit times.

Theorem 5.11 (Exponential convergence of first exit times from $\tilde{D}^{\pm}(\varepsilon^{\gamma})$). *Suppose Hypotheses (H.1) and (H.2) and Convention (C) are satisfied. Then there is a family of random variables $(\bar{\tau}(\varepsilon))_{\varepsilon>0}$ with exponential law of parameter 1 such that for all $\theta < 1$*

$$\lim_{\varepsilon\to 0+} \mathbb{E}\big[\sup_{x\in \hat{D}^{\pm}(\varepsilon^{\gamma})} |\exp\big(\theta\lambda^{\pm}(\varepsilon)\tau_x^{\pm}(\varepsilon)\big) - \exp\big(\theta\bar{\tau}(\varepsilon)\big)|\big] = 0.$$

This implies that the first exit times are of asymptotic order $1/\lambda^{\pm}(\varepsilon) \approx 1/\varepsilon^{\alpha}$ and therefore increase polynomially in the noise parameter as $\varepsilon \to 0+$. This strongly contrasts the behavior known for the Wiener case from [Bra91] and [FJL82], and extends the results of [IP06a, IP06b] and [IP08] to the case of infinite dimensional systems.

Proof. By Theorem 5.8 for each $\theta > -1$ and $C \in (0, 1-\theta)$ there is $\varepsilon_0 > 0$ such that for all $0 < \varepsilon \leqslant \varepsilon_0$

$$\begin{aligned}\frac{1}{1+\theta} - C_1 &\leqslant \mathbb{E}\left[\inf_{x\in \hat{D}^{\pm}(\varepsilon^{\gamma})} \exp\big(-\theta\lambda^{\pm}(\varepsilon)\tau_x^{\pm}(\varepsilon)\big)\right]\\ &\leqslant \mathbb{E}\left[\sup_{x\in \hat{D}^{\pm}(\varepsilon^{\gamma})} \exp\big(-\theta\lambda^{\pm}(\varepsilon)\tau_x^{\pm}(\varepsilon)\big)\right] \leqslant \frac{1}{1+\theta} + C_2.\end{aligned}$$

where $C_1 = \frac{1+C}{1+\theta-C} - \frac{1}{1+\theta} > 0$ and $C_2 = \frac{1}{1}1+\theta - \frac{1-C}{1+\theta+C} > 0$. Clearly $C_1, C_2 \to 0$ for $C \to 0+$. By Theorem 5.10 there is a family of random variables $(\bar{\tau}(\varepsilon))_{\varepsilon>0}$ with exponential law of parameter 1 such that for all $\theta < 1$

$$\mathbb{E}\left[e^{-\theta\bar{\tau}(\varepsilon)}\right] = \frac{1}{1+\theta}.$$

Hence for all $\theta < 1$

$$\mathbb{E}\left[\sup_{x\in \hat{D}^{\pm}(\varepsilon^{\gamma})} \left|\exp\big(\theta\lambda^{\pm}(\varepsilon)\tau_x^{\pm}(\varepsilon)\big) - \exp\big(\theta\bar{\tau}(\varepsilon)\big)\right|\right] \leqslant \max\{C_1, C_2\}.$$

This finishes the proof. □

5.2.1 The Upper Estimate of the Laplace Transform

In this subsection we shall establish the upper estimate part of Theorem 5.8.

Proposition 5.12 (The upper estimate). *Assume that Hypotheses (H.1) and (H.2) and Convention (C) are satisfied. Then for all* $\theta > -1$ *and* $C \in (0, 1+\theta)$ *there exists* $\varepsilon_0 = \varepsilon_0(\theta) > 0$ *such that for all* $0 < \varepsilon \leqslant \varepsilon_0$

$$\mathbb{E}\left[\sup_{x\in\hat{D}^{\pm}(\varepsilon^{\gamma})} \exp\left(-\theta\lambda^{\pm}(\varepsilon)\tau_x^{\pm}(\varepsilon)\right)\right] \leqslant \frac{1+C}{1+\theta-C}.$$

Proof. Fix $\Gamma > 0$ such that Proposition 4.5 and Corollary 4.6 are valid and let C be given as stated. For convenience we drop the superscript $\pm$. Since the jumps of the noise process L exceed any fixed barrier $\mathbb{P}$-a.s., i.e. $\tau_x(\varepsilon)$ is $\mathbb{P}$-a.s. finite, we can rewrite the Laplace transform of $\tau_x(\varepsilon)$ in the following way for $\varepsilon > 0$:

$$\begin{aligned}\mathbb{E}\left[\sup_{x\in\hat{D}(\varepsilon^{\gamma})} e^{-\theta\lambda(\varepsilon)\tau_x(\varepsilon)}\right] = \sum_{k=1}^{\infty}\Bigg(&\mathbb{E}\left[e^{-\theta\lambda(\varepsilon)T_k}\sup_{x\in\hat{D}(\varepsilon^{\gamma})}\mathbf{1}\{\tau_x(\varepsilon)=T_k\}\right]\\ &+\mathbb{E}\left[\sup_{x\in\hat{D}(\varepsilon^{\gamma})} e^{-\theta\lambda(\varepsilon)\tau_x(\varepsilon)}\mathbf{1}\{\tau_x(\varepsilon)\in(T_{k-1},T_k)\}\right]\Bigg).\end{aligned} \tag{5.28}$$

We shall estimate the first and second sum in (5.28) separately. As (5.28) indicates, our arguments will be based on the separation of a large jump compound Poisson part, and a small jump part which does not deviate from the rapidly relaxing deterministic solution trajectories of the Chafee–Infante equation by much. In the second sum, the terms of which are estimated in the following Claims 3 and 4, we are forced to take advantage of the initial values $y \in \hat{D}^{\pm}(\varepsilon^{\gamma})$, while for the first sum initial values in the larger set $\tilde{D}^{\pm}(\varepsilon^{\gamma})$ are shown to be sufficient.

We next estimate the first sum of (5.28)

For $k \in \mathbb{N}$ we can decompose the large jump exit by writing

$$\begin{aligned}&\mathbb{E}\left[e^{-\theta\lambda(\varepsilon)T_k}\sup_{x\in\hat{D}(\varepsilon^{\gamma})}\mathbf{1}\{\tau_x(\varepsilon)=T_k\}\right]\\ &\leqslant \mathbb{E}\left[e^{-\theta\lambda(\varepsilon)T_k}\sup_{x\in\tilde{D}(\varepsilon^{\gamma})}\mathbf{1}\{\tau_x(\varepsilon)=T_k\}\right]\end{aligned}$$

$$= \mathbb{E}\left[e^{-\theta\lambda(\varepsilon)T_k} \sup_{x\in\tilde{D}(\varepsilon^\gamma)} \mathbf{1}\{X^\varepsilon(s;x) \in \tilde{D}(\varepsilon^\gamma) \text{ for } s \in [0,T_k) \text{ and } X^\varepsilon(T_k;x) \notin \tilde{D}(\varepsilon^\gamma)\}\right]$$

$$= \mathbb{E}\left[e^{-\theta\lambda(\varepsilon)T_k} \sup_{x\in\tilde{D}(\varepsilon^\gamma)} \mathbf{1}\left(\bigcap_{i=1}^{k-1} A_{X^\varepsilon(0;x)} \circ \theta_{T_{i-1}} \cap B_{X^\varepsilon(0;x)} \circ \theta_{T_{k-1}}\right)\right]$$

$$= \mathbb{E}\left[e^{-\theta\lambda(\varepsilon)T_k} \sup_{x\in\tilde{D}^\pm(\varepsilon^\gamma)} \prod_{i=1}^{k-1} \mathbf{1}\left(A_{X^\varepsilon(0;x)} \circ \theta_{T_{i-1}}\right) \mathbf{1}\left(B_{X^\varepsilon(0;x)} \circ \theta_{T_{k-1}}\right)\right].$$

Note that $T_k = T_{k-1} + T_1 \circ \theta_{T_{k-1}}$. We use the strong Markov property, conditioning on the past of T_{k-1}, and then estimate from above by the supremum over all values $X(T_{k-1};x)$ can take. This gives

$$\mathbb{E}\left[e^{-\theta\lambda(\varepsilon)T_k} \sup_{x\in\tilde{D}^\pm(\varepsilon^\gamma)} \prod_{i=1}^{k-1} \mathbf{1}\left(A_{X^\varepsilon(0;x)} \circ \theta_{T_{i-1}}\right) \mathbf{1}\left(B_{X^\varepsilon(0;x)} \circ \theta_{T_{k-1}}\right)\right]$$

$$= \mathbb{E}\left[\mathbb{E}\left[e^{-\theta\lambda(\varepsilon)T_k} \sup_{x\in\tilde{D}^\pm(\varepsilon^\gamma)} \prod_{i=1}^{k-1} \mathbf{1}\left(A_{X^\varepsilon(0;x)\circ\theta_{T_{i-1}}}\right) \mathbf{1}\left(B_{X^\varepsilon(0;x)} \circ \theta_{T_{k-1}}\right) | \mathscr{F}_{T_{k-1}}\right]\right]$$

$$\leqslant \mathbb{E}\left[e^{-\theta\lambda(\varepsilon)T_{k-1}} \sup_{x\in\tilde{D}^\pm(\varepsilon^\gamma)} \prod_{i=1}^{k-2} \mathbf{1}\left(A_{X^\varepsilon(0;x)} \circ \theta_{T_{i-1}}\right) \mathbf{1}\left(A_{X^\varepsilon(0;x)} \circ \theta_{T_{k-2}}\right)\right]$$

$$\cdot \mathbb{E}\left[e^{-\theta\lambda(\varepsilon)\, T_1} \sup_{y\in\tilde{D}(\varepsilon^\gamma)} \mathbf{1}\left(B_y\right)\right].$$

By $k-1$-fold iteration of this argument we obtain for $k \in \mathbb{N}$

$$\mathbb{E}\left[\sup_{x\in\tilde{D}(\varepsilon^\gamma)} e^{-\theta\lambda(\varepsilon)T_k} \mathbf{1}\{\tau_x(\varepsilon) = T_k\}\right]$$

$$\leqslant \left(\mathbb{E}\left[e^{-\theta\lambda(\varepsilon)T_1} \sup_{y\in\tilde{D}(\varepsilon^\gamma)} \mathbf{1}\left(A_y\right)\right]\right)^{k-1} \mathbb{E}\left[e^{-\theta\lambda(\varepsilon)T_1} \sup_{y\in\tilde{D}(\varepsilon^\gamma)} \mathbf{1}\left(B_y\right)\right].$$

Now we have to estimate the individual terms corresponding to A_y and B_y by exploiting Lemma 5.2.

Claim 1. There exists $\varepsilon_0 > 0$ such that for all $0 < \varepsilon \leqslant \varepsilon_0$

$$\mathbb{E}_x\left[e^{-\theta\lambda(\varepsilon)T_1} \sup_{y\in\tilde{D}(\varepsilon^\gamma)} \mathbf{1}(A_y)\right] \leqslant \frac{\beta_\varepsilon}{\beta_\varepsilon + \theta\lambda(\varepsilon)} \left(1 - \frac{\lambda(\varepsilon)}{\beta_\varepsilon}(1 - \frac{C}{5})\right).$$

In the inequality of Lemma 5.2 (*ix*) we can pass to the supremum in $y \in \tilde{D}(\varepsilon^\gamma)$, and integrate to obtain, using the independence of jump times and heights

$$
\begin{aligned}
&\mathbb{E}\left[e^{-\theta\lambda(\varepsilon)T_1} \sup_{y\in\tilde{D}(\varepsilon^\gamma)} \mathbf{1}(A_y)\right] \\
&\leqslant \mathbb{E}\left[e^{-\theta\lambda(\varepsilon)T_1}\mathbf{1}\{T_1 < s_{r^*} + T_{rec} + \kappa\gamma|\ln\varepsilon|\}\right]\mathbb{P}\left(\varepsilon\|W_1\| \geqslant (1/2)\varepsilon^{2\gamma}\right) \\
&\quad + \mathbb{E}\left[e^{-\theta\lambda(\varepsilon)T_1}\right]\mathbb{P}\left(\varepsilon W_1 \in D_0\right) + \mathbb{E}\left[e^{-\theta\lambda(\varepsilon)T_1} \sup_{y\in\tilde{D}(\varepsilon^\gamma)} \mathbf{1}(E_y^c)\right] \\
&=: K_1K_2 + K_3K_4 + K_5.
\end{aligned}
$$

Let us estimate $K_1, \ldots, K_5$ separately, in the order of increasing complexity. We shall see that the asymptotic behavior of the aggregate is dominated by the second summand.

1. Clearly

$$
K_3 = \int_0^\infty e^{-\theta\lambda(\varepsilon)s}\beta_\varepsilon e^{-\beta_\varepsilon s}\,\mathrm{d}s = \frac{\beta_\varepsilon}{\beta_\varepsilon + \theta\lambda(\varepsilon)}. \tag{5.29}
$$

2. By definition of $\lambda(\varepsilon)$ we know

$$
K_4 = \mathbb{P}\left(W_1 \in (1/\varepsilon)D_0\right) = 1 - \lambda(\varepsilon)/\beta_\varepsilon. \tag{5.30}
$$

3. The recurrence time of logarithmic order in ε enters into the calculation of K_1. Remember that κ is fixed along with Γ. We have

$$
\begin{aligned}
K_1 &= \int_0^{s_{r^*} + T_{rec} + \kappa\gamma|\ln\varepsilon|} e^{-\theta\lambda(\varepsilon)s}\beta_\varepsilon e^{-\beta_\varepsilon s}\,\mathrm{d}s \\
&= \frac{\beta_\varepsilon}{\theta\lambda(\varepsilon) + \beta_\varepsilon}\left[1 - \exp\left(-(\theta\lambda(\varepsilon) + \beta_\varepsilon)(s_{r^*} + T_{rec} + \kappa\gamma|\ln\varepsilon|)\right)\right].
\end{aligned}
$$

4. For the estimation of $K_2 = \mathbb{P}\left(\|\varepsilon W_1\| \geqslant (1/2)\varepsilon^{2\gamma}\right)$ we use Lemma 5.5 (*II*), providing, for any $a > 0$, $\varepsilon_0 > 0$ such that for $0 < \varepsilon \leqslant \varepsilon_0$

$$
K_2 \leqslant \varepsilon^{\alpha(1-2\gamma-\rho)-a}. \tag{5.31}
$$

5. For K_5 we refer to Corollary 4.6 ensuring that for ε_0 small enough we have for $0 < \varepsilon \leqslant \varepsilon_0$

$$K_5 \leqslant \frac{C}{10} \frac{\beta_\varepsilon}{\beta_\varepsilon + \theta\lambda(\varepsilon)} \frac{\lambda(\varepsilon)}{\beta_\varepsilon}. \tag{5.32}$$

Inserting the estimates we obtained for $K_1, \ldots, K_5$ into our original inequality we can write for ε_0 small enough and $0 < \varepsilon \le \varepsilon_0$

$$\begin{aligned}
&\mathbb{E}\left[e^{-\theta\lambda(\varepsilon)T_1} \sup_{y \in \tilde{D}(\varepsilon^\gamma)} \mathbf{1}(A_y)\right] \\
&\leqslant \frac{\beta_\varepsilon}{\beta_\varepsilon + \theta\lambda(\varepsilon)} \left[1 - \exp\left(-(\theta\lambda(\varepsilon) + \beta_\varepsilon)(s_{r^*} + T_{rec} + \kappa\gamma|\ln\varepsilon|)\right)\right] \varepsilon^{\alpha(1-2\gamma-\rho)-a} \\
&\quad + \frac{\beta_\varepsilon}{\beta_\varepsilon + \theta\lambda(\varepsilon)} \left(1 - \frac{\lambda(\varepsilon)}{\beta_\varepsilon}\right) + \frac{C}{10} \frac{\beta_\varepsilon}{\beta_\varepsilon + \theta\lambda(\varepsilon)} \frac{\lambda(\varepsilon)}{\beta_\varepsilon} \\
&= \frac{\beta_\varepsilon}{\beta_\varepsilon + \theta\lambda(\varepsilon)} \left[1 - \frac{\lambda(\varepsilon)}{\beta_\varepsilon}(1 - \frac{C}{10})\right] \\
&\quad + \frac{\beta_\varepsilon}{\beta_\varepsilon + \theta\lambda(\varepsilon)} \left[1 - \exp\left(-(\theta\lambda(\varepsilon) + \beta_\varepsilon)(s_{r^*} + T_{rec} + \kappa\gamma|\ln\varepsilon|)\right)\right] 4\varepsilon^{\alpha(1-2\gamma-\rho)} \\
&= \frac{\beta_\varepsilon}{\beta_\varepsilon + \theta\lambda(\varepsilon)} \left[1 - \frac{\lambda(\varepsilon)}{\beta_\varepsilon}\left(1 - \frac{C}{10} - K_6\right)\right],
\end{aligned}$$

where

$$K_6 = \frac{\beta_\varepsilon}{\lambda(\varepsilon)} \left[1 - \exp\left(-(\theta\lambda(\varepsilon) + \beta_\varepsilon)(s_{r^*} + T_{rec} + \kappa\gamma|\ln\varepsilon|)\right)\right] \varepsilon^{\alpha(1-2\gamma-\rho)-a}.$$

To estimate K_6, recall that by the equalities (3.29) and (3.30) and Lemma 5.5 (I) there exists $\varepsilon_0 > 0$ such that for $0 < \varepsilon \leqslant \varepsilon_0$

$$\frac{\beta_\varepsilon}{\lambda(\varepsilon)} \leqslant \varepsilon^{-\alpha(1-\rho)+a}. \tag{5.33}$$

By (5.33) and $2\gamma < \rho$ we know that by eventually choosing a and ε_0 sufficiently small we may obtain for $0 < \varepsilon \leqslant \varepsilon_0$

$$K_6 \leqslant (\theta\lambda(\varepsilon) + \beta_\varepsilon)(s_{r^*} + T_{rec} + \kappa\gamma|\ln\varepsilon|)\varepsilon^{-2\alpha\gamma - as} \leqslant \frac{C}{10}.$$

We can summarize our findings in stating that there exists $\varepsilon_0 > 0$ such that for $0 < \varepsilon \leqslant \varepsilon_0$

$$\mathbb{E}\left[e^{-\theta\lambda(\varepsilon)T_1} \sup_{x \in \tilde{D}(\varepsilon^\gamma)} \mathbf{1}(A_x)\right] \leqslant \frac{\beta_\varepsilon}{\beta_\varepsilon + \theta\lambda(\varepsilon)} \left(1 - \frac{\lambda(\varepsilon)}{\beta_\varepsilon}\left(1 - \frac{C}{5}\right)\right). \tag{5.34}$$

Claim 2. There is $\varepsilon_0 > 0$ such that for all $0 < \varepsilon \leqslant \varepsilon_0$

$$\mathbb{E}\left[e^{-\theta\lambda(\varepsilon)T_1} \sup_{y\in\tilde{D}(\varepsilon^\gamma)} \mathbf{1}\left(B_y\right)\right] \leqslant (1+C)\frac{\beta_\varepsilon}{\beta_\varepsilon+\theta\lambda(\varepsilon)}\frac{\lambda(\varepsilon)}{\beta_\varepsilon}.$$

In the inequality of Lemma 5.2 (x) we again pass to the supremum in $y \in \tilde{D}(\varepsilon^\gamma)$, and integrate to obtain, using the independence of jump times and height increments

$$\begin{aligned}
&\mathbb{E}\left[e^{-\theta\lambda(\varepsilon)T_1} \sup_{y\in\tilde{D}(\varepsilon^\gamma)} \mathbf{1}\left(B_y\right)\right]\\
&\leqslant \mathbb{E}\left[e^{-\theta\lambda(\varepsilon)T_1}\right]\mathbb{P}\left(\varepsilon W_1 \in (\hat{D}_0(\varepsilon^\gamma))^c\right)\\
&\quad + \mathbb{E}\left[e^{-\theta\lambda(\varepsilon)T_1}\mathbf{1}\{T_1 < s_{r*} + T_{rec} + \kappa\gamma|\ln\varepsilon|\}\right] + \mathbb{E}\left[e^{-\theta\lambda(\varepsilon)T_1} \sup_{y\in\tilde{D}(\varepsilon^\gamma)} \mathbf{1}(E_y^c)\right]\\
&=: K_3K_8 + K_1 + K_5.
\end{aligned} \tag{5.35}$$

Examining K_1 more closely, we recognize by Lemma 5.5 (I) and by $\rho > 1/2$ that there exists $\varepsilon_0 > 0$ such that for $0 < \varepsilon \leqslant \varepsilon_0$ we have

$$\begin{aligned}
K_1 &= \frac{\beta_\varepsilon}{\theta\lambda(\varepsilon)+\beta_\varepsilon}\left[1-\exp\left(-(\theta\lambda(\varepsilon)+\beta_\varepsilon)(s_{r*}+T_{rec}+\kappa\gamma|\ln\varepsilon|)\right)\right]\\
&\leqslant \frac{\beta_\varepsilon}{\theta\lambda(\varepsilon)+\beta_\varepsilon}(\theta\lambda(\varepsilon)+\beta_\varepsilon)(s_{r*}+T_{rec}+\kappa\gamma|\ln\varepsilon|)\\
&\leqslant \frac{\beta_\varepsilon}{\theta\lambda(\varepsilon)+\beta_\varepsilon}\frac{\lambda(\varepsilon)}{\beta_\varepsilon}\left(\frac{\beta_\varepsilon}{\lambda(\varepsilon)}(\theta\lambda(\varepsilon)+\beta_\varepsilon)(s_{r*}+T_{rec}+\kappa\gamma|\ln\varepsilon|)\right)\\
&\leqslant \frac{\beta_\varepsilon}{\theta\lambda(\varepsilon)+\beta_\varepsilon}\frac{\lambda(\varepsilon)}{\beta_\varepsilon}\frac{C}{10}.
\end{aligned} \tag{5.36}$$

In order to estimate K_8 we use Lemma 5.5 (VI). It yields that there is $\varepsilon_0 > 0$ such that for $0 < \varepsilon \leqslant \varepsilon_0$

$$K_8 \leqslant (1+\frac{C}{5})\frac{\lambda(\varepsilon)}{\beta_\varepsilon}.$$

Recalling the estimates for K_3 and K_5 from the preceding part, we find ε_0 such that for $0 < \varepsilon \leqslant \varepsilon_0$

$$\mathbb{E}\left[e^{-\theta\lambda(\varepsilon)T_1} \sup_{y\in\tilde{D}(\varepsilon^\gamma)} \mathbf{1}(B(y))\right] \leqslant (1+C)\frac{\beta_\varepsilon}{\beta_\varepsilon+\theta\lambda(\varepsilon)}\frac{\lambda(\varepsilon)}{\beta_\varepsilon}.$$

We now estimate the second sum of (5.28)

In order to treat the summands

$$\mathbb{E}\left[\sup_{x\in\hat{D}(\varepsilon^\gamma)} e^{-\theta\lambda(\varepsilon)\tau_x(\varepsilon)}\mathbf{1}\{\tau_x(\varepsilon)\in(T_{k-1},T_k)\}\right]$$

for $k\in\mathbb{N}$ we have to distinguish the cases $\theta\geqslant 0$ and $\theta\in(-1,0)$. More precisely, for $k\in\mathbb{N}$ we start with the inequality

$$\mathbb{E}\left[\sup_{x\in\hat{D}(\varepsilon^\gamma)} e^{-\theta\lambda(\varepsilon)\tau_x(\varepsilon)}\mathbf{1}\{\tau_x(\varepsilon)\in(T_{k-1},T_k)\}\right]$$
$$\leqslant\begin{cases}\mathbb{E}\left[e^{-\theta\lambda(\varepsilon)T_{k-1}}\sup_{x\in\hat{D}(\varepsilon^\gamma)}\mathbf{1}\{\tau_x(\varepsilon)\in(T_{k-1},T_k)\}\right], & \text{if } \theta\geqslant 0,\\ \mathbb{E}\left[e^{-\theta\lambda(\varepsilon)T_{k}}\sup_{x\in\hat{D}(\varepsilon^\gamma)}\mathbf{1}\{\tau_x(\varepsilon)\in(T_{k-1},T_k)\}\right], & \text{if } -1<\theta<0.\end{cases}$$

We have to argue separately for the cases $k=1$ and $k\geqslant 2$. It is the case $k=1$, in which we need to take advantage of the initial value $x\in\hat{D}(\varepsilon^\gamma)$.

Claim 3. There is $\varepsilon_0>0$ such that for all $0<\varepsilon\leqslant\varepsilon_0$

$$\mathbb{E}\left[\sup_{x\in\hat{D}(\varepsilon^\gamma)} e^{-\theta\lambda(\varepsilon)\tau_x(\varepsilon)}\mathbf{1}\{\tau_x(\varepsilon)\in(0,T_1)\}\right]\leqslant\frac{C}{5}\left(\frac{\beta_\varepsilon}{\beta_\varepsilon+\theta\lambda(\varepsilon)}\right)\frac{\lambda(\varepsilon)}{\beta_\varepsilon}.$$

1. First consider the case $\theta\geqslant 0$. With $T_0=0$ we see with the help of Lemma 5.2 (xi)

$$\mathbb{E}\left[\sup_{x\in\hat{D}(\varepsilon^\gamma)} e^{-\theta\lambda(\varepsilon)T_0}\mathbf{1}\{\tau_x(\varepsilon)\in(0,T_1)\}\right]$$
$$\leqslant\mathbb{E}\left[\sup_{y\in\hat{D}(\varepsilon^\gamma)}\mathbf{1}\{Y^\varepsilon(s;y)\notin\tilde{D}(\varepsilon^\gamma)\text{ for some } s\in(0,T_1)\}\right]$$
$$\leqslant\mathbb{E}\left[\sup_{y\in\hat{D}(\varepsilon^\gamma)}\mathbf{1}(E_y^c)\right]+\mathbb{P}(T_1<s_{r^*}+T_{rec}+\kappa\gamma|\ln\varepsilon|)$$
$$\leqslant\mathbb{E}\left[\sup_{y\in\tilde{D}(\varepsilon^\gamma)}\mathbf{1}(E_y^c)\right]+\mathbb{P}(T_1<s_{r^*}+T_{rec}+\kappa\gamma|\ln\varepsilon|).$$

We apply Corollary 4.6 and a similar estimate as for the term K_1 in the proof of Claim 1, which guarantee that there is $\varepsilon_0>0$ such that for $0<\varepsilon\leqslant\varepsilon_0$

$$\mathbb{E}\left[\sup_{x\in\hat{D}(\varepsilon^\gamma)} e^{-\theta\lambda(\varepsilon)T_0}\mathbf{1}\{\tau_x(\varepsilon)\in(T_0,T_1)\}\right]\leqslant\frac{C}{5}\frac{\beta_\varepsilon}{\beta_\varepsilon+\theta\lambda(\varepsilon)}\frac{\lambda(\varepsilon)}{\beta_\varepsilon}.$$

2. In case $\theta\in(-1,0)$ we may write, thanks to Lemma 5.2 (xi),

$$\mathbb{E}\left[\sup_{x\in\hat{D}(\varepsilon^\gamma)} e^{-\theta\lambda(\varepsilon)\tau_x(\varepsilon)}\mathbf{1}\{\tau_x(\varepsilon)\in(0,T_1)\}\right]$$

$$\leqslant\mathbb{E}\left[e^{-\theta\lambda(\varepsilon)T_1}\sup_{x\in\hat{D}(\varepsilon^\gamma)}\mathbf{1}\{X^\varepsilon(s;x)\notin\tilde{D}(\varepsilon^\gamma)\text{ for some }s\in(0,T_1)\}\right]$$

$$\leqslant\mathbb{E}\left[e^{-\theta\lambda(\varepsilon)T_1}\sup_{y\in\hat{D}(\varepsilon^\gamma)}\mathbf{1}(E_y^c)\right]+\mathbb{E}\left[e^{-\theta\lambda(\varepsilon)T_1}\mathbf{1}\{T_1<s_{r*}+T_{rec}+\kappa\gamma|\ln\varepsilon|\}\right]$$

$$\leqslant\mathbb{E}\left[e^{-\theta\lambda(\varepsilon)T_1}\sup_{y\in\tilde{D}(\varepsilon^\gamma)}\mathbf{1}(E_y^c)\right]+\mathbb{E}\left[e^{-\theta\lambda(\varepsilon)T_1}\mathbf{1}\{T_1<s_{r*}+T_{rec}+\kappa\gamma|\ln\varepsilon|\}\right].$$

Using Corollary 4.6 and the estimate of the term K_1 in the proof of Claim 1, we obtain for sufficiently small $\varepsilon>0$ an analogous inequality

$$\mathbb{E}\left[\sup_{x\in\hat{D}(\varepsilon^\gamma)} e^{-\theta\lambda(\varepsilon)T_1}\mathbf{1}\{\tau_x(\varepsilon)\in(0,T_1)\}\right]=\left(\frac{\beta_\varepsilon}{\beta_\varepsilon+\theta\lambda(\varepsilon)}\right)\frac{\lambda(\varepsilon)}{\beta_\varepsilon}\frac{C}{5}. \tag{5.37}$$

We continue for the case $k\geqslant 2$.

Claim 4. There exists $\varepsilon_0>0$ such that for any $k\geqslant 2$ and $0<\varepsilon\leqslant\varepsilon_0$

$$\mathbb{E}\left[\sup_{x\in\hat{D}(\varepsilon^\gamma)} e^{-\theta\lambda(\varepsilon)\tau_x(\varepsilon)}\mathbf{1}\{\tau_x(\varepsilon)\in(T_{k-1},T_k)\}\right]$$
$$\leqslant\left(\frac{\beta_\varepsilon}{\beta_\varepsilon+\theta\lambda(\varepsilon)}\left(1-\frac{\lambda(\varepsilon)}{\beta_\varepsilon}\left(1-\frac{C}{5}\right)\right)\right)^{k-2}\frac{C}{5}\frac{\beta_\varepsilon}{\beta_\varepsilon+\theta\lambda(\varepsilon)}\frac{\lambda(\varepsilon)}{\beta_\varepsilon}.$$

1. For $\theta\geqslant 0$ we use the strong Markov property as in the estimate for the first summand to get for $k\geqslant 2$

$$\mathbb{E}\left[\sup_{x\in\hat{D}(\varepsilon^\gamma)} e^{-\theta\lambda(\varepsilon)T_{k-1}}\mathbf{1}\{\tau_x(\varepsilon)\in(T_{k-1},T_k)\}\right]$$
$$\leqslant\left(\mathbb{E}\left[e^{-\theta\lambda(\varepsilon)T_1}\sup_{y\in\tilde{D}(\varepsilon^\gamma)}\mathbf{1}(A_y)\right]\right)^{k-2}$$

$$\cdot\, \mathbb{E}\left[e^{-\theta\lambda(\varepsilon)T_1} \sup_{y\in\tilde{D}(\varepsilon^\gamma)} \mathbf{1}(A_y)\mathbf{1}\{\exists s \in (0,t_2),\ Y^\varepsilon(s;X^\varepsilon(0;y))\circ\theta_{T_1} \notin \tilde{D}(\varepsilon^\gamma)\}\right]. \tag{5.38}$$

The event appearing in the last integral is estimated in Lemma 5.2 (*xii*). With this in mind we obtain

$$\begin{aligned}
&\mathbb{E}\left[e^{-\theta\lambda(\varepsilon)T_1} \sup_{y\in\tilde{D}(\varepsilon^\gamma)} \mathbf{1}(A_y)\mathbf{1}\{Y^\varepsilon(s,X^\varepsilon(0;y))\circ\theta_{T_1} \notin \tilde{D}(\varepsilon^\gamma) \text{ f. s. } s \in (0,t_2)\}\right]\\
&\leqslant \mathbb{E}\left[e^{-\theta\lambda(\varepsilon)T_1}\right]\mathbb{P}\left(\varepsilon W_1 \in D_0^*(\varepsilon^\gamma)\right) + \mathbb{E}\left[e^{-\theta\lambda(\varepsilon)T_1}\mathbf{1}\{T_1 < s_{r^*}+T_{rec}+\kappa\gamma|\ln\varepsilon|\}\right]\\
&\quad + \mathbb{E}\left[e^{-\theta\lambda(\varepsilon)T_1}\mathbf{1}\{T_1 < s_{r^*}+T_{rec}+\kappa\gamma|\ln\varepsilon|\}\circ\theta_{T_1}\right] + 2\,\mathbb{E}\left[\sup_{y\in\tilde{D}(\varepsilon^\gamma)}\mathbf{1}(E_y^c)\right]\\
&=: K_3K_9 + K_1 + \tilde{K}_1 + 2K_5.
\end{aligned}$$

Lemma 5.5 (*IV*) provides $\varepsilon_0 > 0$ such that for $0 < \varepsilon \leqslant \varepsilon_0$

$$K_9 = \mathbb{P}\left(\varepsilon W_1 \in D_0^*(\varepsilon^\gamma)\right) \leqslant \frac{C}{20}\frac{\lambda(\varepsilon)}{\beta_\varepsilon}.$$

The asymptotic behavior of K_3, K_5 and K_1 as ε tends to 0 is known from previous parts of the proof and $\tilde{K}_1$ is estimated similarly as K_1. We may therefore deduce that there is $\varepsilon_0 > 0$ such that for all $0 < \varepsilon \leqslant \varepsilon_0$

$$\begin{aligned}
&\mathbb{E}\left[e^{-\theta\lambda(\varepsilon)T_1} \sup_{y\in\tilde{D}(\varepsilon^\gamma)} \mathbf{1}(A_y)\mathbf{1}\{Y^\varepsilon(s,X^\varepsilon(0;x))\circ\theta_{T_1} \notin \tilde{D}(\varepsilon^\gamma) \text{ for s. } s \in (0,t_2)\}\right]\\
&\leqslant \frac{C}{5}\frac{\beta_\varepsilon}{\beta_\varepsilon+\theta\lambda(\varepsilon)}\frac{\lambda(\varepsilon)}{\beta_\varepsilon}.
\end{aligned} \tag{5.39}$$

An estimate for the first factor in (5.34) is known from Claim 1. This completes the proof of the inequality of Claim 4 for $\theta \geqslant 0$.

2. Let us consider the case $\theta \in (-1,0)$. With arguments as before employing the strong Markov property we obtain this time the estimate

$$\mathbb{E}\left[e^{-\theta\lambda(\varepsilon)T_{k-1}} \sup_{x\in\hat{D}(\varepsilon^\gamma)} \mathbf{1}\{\tau_x(\varepsilon) \in (T_{k-1},T_k)\}\right]$$

$$\leqslant \left(\mathbb{E}\left[e^{-\theta\lambda(\varepsilon)T_1} \sup_{y\in\tilde{D}(\varepsilon^\gamma)} \mathbf{1}(A_y) \right] \right)^{k-2}$$

$$\cdot\, \mathbb{E}\left[e^{-\theta\lambda(\varepsilon)(T_1+t_2\circ\theta_{T_1})} \sup_{y\in\tilde{D}(\varepsilon^\gamma)} \mathbf{1}(A_y)\mathbf{1}\{Y^\varepsilon(X^\varepsilon(0,y))\circ\theta_{T_1} \notin \tilde{D}(\varepsilon^\gamma) \text{ f.s } s \in (0,T_1)\} \right].$$

To estimate the last factor in the previous expression, we use the strong Markov property once again, and then Claim 1 and the previous result, keeping in mind Lemma 5.2. Since $T_2 = T_1 + t_2 \circ \theta_{T_1}$ we infer similarly as before but with the full strength of Corollary 4.6 that there is $\varepsilon_0 > 0$ such that for all $0 < \varepsilon \leqslant \varepsilon_0$

$$\mathbb{E}\left[e^{-\theta\lambda(\varepsilon)T_2} \sup_{y\in\tilde{D}(\varepsilon^\gamma)} \mathbf{1}(A_y)\mathbf{1}\{Y^\varepsilon(X^\varepsilon(0,y))\circ\theta_{T_1} \notin \tilde{D}(\varepsilon^\gamma) \text{ for s. } s \in (0,T_1)\} \right]$$
$$\leqslant \frac{C}{5}\frac{\beta_\varepsilon}{\beta_\varepsilon+\theta\lambda(\varepsilon)}\frac{\lambda(\varepsilon)}{\beta_\varepsilon}.$$

This provides the same estimate (5.39) as in the case $\theta \geqslant 0$ and completes the proof of Claim 4.

Combining the estimates for the first and second summands in (5.28) by Claims 1–4 we find $\varepsilon_0 > 0$ such that for all $0 < \varepsilon \leqslant \varepsilon_0$

$$\mathbb{E}\left[\sup_{x\in\hat{D}(\varepsilon^\gamma)} e^{-\theta\lambda(\varepsilon)\tau_x(\varepsilon)} \right]$$
$$\leqslant \sum_{k=1}^{\infty} \left(\frac{\beta_\varepsilon}{\beta_\varepsilon+\theta\lambda(\varepsilon)}\right)^{k-1} \left(1-\frac{\lambda(\varepsilon)}{\beta_\varepsilon}(1-\frac{C}{5})\right)^{k-1} \frac{\beta_\varepsilon}{\beta_\varepsilon+\theta\lambda(\varepsilon)}\frac{\lambda(\varepsilon)}{\beta_\varepsilon}(1+\frac{C}{5})$$
$$+\frac{C}{5}\frac{\beta_\varepsilon}{\beta_\varepsilon+\theta\lambda(\varepsilon)}\frac{\lambda(\varepsilon)}{\beta_\varepsilon}$$
$$+\sum_{k=2}^{\infty} \left(\frac{\beta_\varepsilon}{\beta_\varepsilon+\theta\lambda(\varepsilon)}\right)^{k-2} \left(1-\frac{\lambda(\varepsilon)}{\beta_\varepsilon}(1-\frac{C}{5})\right)^{k-2} \frac{C}{5}\frac{\beta_\varepsilon}{\beta_\varepsilon+\theta\lambda(\varepsilon)}\frac{\lambda(\varepsilon)}{\beta_\varepsilon}$$
$$\leqslant \left(1+\frac{2C}{5}\right)\frac{\lambda(\varepsilon)}{\beta_\varepsilon}\frac{\beta_\varepsilon}{\beta_\varepsilon+\theta\lambda(\varepsilon)}\sum_{k=0}^{\infty}\left(\frac{\beta_\varepsilon}{\beta_\varepsilon+\theta\lambda(\varepsilon)}\left(1-\frac{\lambda(\varepsilon)}{\beta_\varepsilon}(1-\frac{C}{5})\right)\right)^k$$
$$\leqslant \frac{1+C}{\theta+(1-C)}.$$

The sum obviously converges if and only if $C < \theta + 1$. □

5.2.2 The Lower Estimate of the Laplace Transform

The lower estimate is easier to obtain since we can neglect the non-negative second sum in (5.28). The tedious reasoning concerning small deviations of the small noise part from the deterministic solution trajectories of the Chafee–Infante equation is not needed.

Proposition 5.13 (The lower estimate). *Assume that Hypothesis (H.1) and (H.2) and Convention (C) are satisfied. Then for all $\theta > -1$ and $C \in (0, 1+\theta)$ there is $\varepsilon_0 = \varepsilon_0(\theta) > 0$ such that for all $0 < \varepsilon \leqslant \varepsilon_0$*

$$\mathbb{E}\left[\inf_{x\in\hat{D}^{\pm}(\varepsilon^{\gamma})} \exp\left(-\theta\lambda^{\pm}(\varepsilon)\tau_x^{\pm}(\varepsilon)\right)\right] \geqslant \frac{1+C}{1+\theta-C}.$$

Proof. Again we omit the superscript $\pm$ and fix $\Gamma > 0$ large enough such that Proposition 4.5 and Corollary 4.6 are true. Reducing equation (5.28) in the way indicated, and applying the strong Markov property in the usual way we obtain the estimate

$$\begin{aligned}
&\mathbb{E}\left[\inf_{x\in\hat{D}(\varepsilon^{\gamma})} e^{-\theta\lambda(\varepsilon)\tau_x(\varepsilon)}\right] \\
&\geqslant \sum_{k=1}^{\infty} \mathbb{E}\left[\inf_{x\in\hat{D}(\varepsilon^{\gamma})} e^{-\theta\lambda(\varepsilon)T_k}\mathbf{1}\{\tau_x(\varepsilon) = T_k\}\right] \\
&= \sum_{k=1}^{\infty} \mathbb{E}\left[e^{-\theta\lambda(\varepsilon)T_k} \inf_{x\in\hat{D}(\varepsilon^{\gamma})} \mathbf{1}\{X^{\varepsilon}(s;x) \in \tilde{D}(\varepsilon^{\gamma}), s\in[0,T_k), X^{\varepsilon}(T_k;x)\notin \tilde{D}(\varepsilon^{\gamma})\}\right] \\
&\geqslant \sum_{k=1}^{\infty} \mathbb{E}\left[e^{-\theta\lambda(\varepsilon)T_k} \inf_{x\in\hat{D}(\varepsilon^{\gamma})} \mathbf{1}\left(\bigcap_{i=1}^{k-1} A^{-}_{X^{\varepsilon}(0;x))}\circ\theta_{T_{i-1}} \cap B_{X^{\varepsilon}(0;x))}\circ\theta_{T_{k-1}}\right)\right] \\
&\geqslant \sum_{k=1}^{\infty} \left(\mathbb{E}\left[e^{-\theta\lambda(\varepsilon)T_1} \inf_{y\in\hat{D}(\varepsilon^{\gamma})} \mathbf{1}(A_y^{-})\right]\right)^{k-1} \mathbb{E}\left[e^{-\theta\lambda(\varepsilon)T_1} \inf_{y\in\hat{D}(\varepsilon^{\gamma})} \mathbf{1}(B_y)\right]. \qquad (5.40)
\end{aligned}$$

Let us treat the terms appearing in (5.40) in a similar way as for the upper estimate.

Claim 1. There is $\varepsilon_0 > 0$ such that for all $0 < \varepsilon \leqslant \varepsilon_0$

$$\mathbb{E}\left[e^{-\theta\lambda(\varepsilon)T_1} \inf_{x\in\hat{D}(\varepsilon^{\gamma})} \mathbf{1}(A_x^{-})\right] \geqslant \frac{\beta_\varepsilon}{\beta_\varepsilon + \theta\lambda(\varepsilon)}\left(1-(1+C)\frac{\lambda(\varepsilon)}{\beta_\varepsilon}\right).$$

First we apply Lemma 5.2 (*xiii*) and take the infimum over $y \in \hat{D}(\varepsilon^{\gamma})$ and integrate, to obtain

$$
\begin{aligned}
&\mathbb{E}\left[e^{-\theta\lambda(\varepsilon)T_1}\inf_{y\in\hat{D}(\varepsilon^\gamma)}\mathbf{1}(A_y^-)\right]\\
&\quad\geqslant \mathbb{E}\left[\varepsilon^{-\theta\lambda(\varepsilon)T_1}\right]\mathbb{P}\left(\varepsilon W_1\in D_0(\varepsilon^\gamma,\varepsilon^{2\gamma},\varepsilon^{2\gamma},\varepsilon^{2\gamma})\right)\\
&\quad\quad-\mathbb{E}\left[e^{-\theta\lambda(\varepsilon)T_1}\mathbf{1}\{T_1<s_{r^*}+T_{rec}+\kappa\gamma|\ln\varepsilon|\}\right]\\
&\quad\quad-\mathbb{E}\left[e^{-\theta\lambda(\varepsilon)T_1}\sup_{y\in\hat{D}(\varepsilon^\gamma)}\mathbf{1}(E^c(y))\right]\\
&\quad=K_3\left(1-\mathbb{P}(W_1\in(1/\varepsilon)(D_0(\varepsilon^\gamma,\varepsilon^{2\gamma},\varepsilon^{2\gamma},\varepsilon^{2\gamma}))^c\right)-K_1-K_5.
\end{aligned}
$$

By Lemma 5.5 (V) there exists $\varepsilon_0>0$ such that for $0<\varepsilon\leqslant\varepsilon_0$

$$
\mathbb{P}(\varepsilon W_1\in D_0^c(\varepsilon^\gamma,\varepsilon^{2\gamma},\varepsilon^{2\gamma},\varepsilon^{2\gamma}))\leqslant(1+\frac{C}{5})\frac{\lambda(\varepsilon)}{\beta_\varepsilon}.
$$

Using the estimates for K_1, K_3, K_5 derived in the proof of the upper estimate, we finally find $\varepsilon_0>0$ such that for $0<\varepsilon\leqslant\varepsilon_0$

$$
\begin{aligned}
&\mathbb{E}\left[e^{-\theta\lambda(\varepsilon)T_1}\inf_{x\in\hat{D}(\varepsilon^\gamma)}\mathbf{1}(A_x^-)\right]\\
&\quad\geqslant\frac{\beta_\varepsilon}{\beta_\varepsilon+\theta\lambda(\varepsilon)}\left(1-(1+\frac{C}{5})\frac{\lambda(\varepsilon)}{\beta_\varepsilon}\right)-\frac{C}{5}\frac{\beta_\varepsilon}{\beta_\varepsilon+\theta\lambda(\varepsilon)}\frac{\lambda(\varepsilon)}{\beta_\varepsilon}\\
&\quad\quad-\frac{C}{5}\frac{\beta_\varepsilon}{\beta_\varepsilon+\theta\lambda(\varepsilon)}\frac{\lambda(\varepsilon)}{\beta_\varepsilon}-\frac{2C}{5}\frac{\beta_\varepsilon}{\theta\lambda(\varepsilon)+\beta_\varepsilon}\frac{\lambda(\varepsilon)}{\beta_\varepsilon}\\
&\quad\geqslant\frac{\beta_\varepsilon}{\beta_\varepsilon+\theta\lambda(\varepsilon)}\left(1-(1+C)\frac{\lambda(\varepsilon)}{\beta_\varepsilon}\right).
\end{aligned}
$$

Claim 2. There is $\varepsilon_0>0$ such that for $0<\varepsilon\leqslant\varepsilon_0$

$$
\mathbb{E}\left[e^{-\theta\lambda(\varepsilon)T_1}\inf_{y\in\hat{D}(\varepsilon^\gamma)}\mathbf{1}(B_y)\right]\geqslant\frac{\beta_\varepsilon}{\theta\lambda(\varepsilon)+\beta_\varepsilon}\left((1-C)\frac{\lambda(\varepsilon)}{\beta_\varepsilon}\right).
$$

Using Lemma 5.2 (xiv) we can infer that

$$
\begin{aligned}
&\mathbb{E}\left[e^{-\theta\lambda(\varepsilon)T_1}\inf_{y\in\hat{D}(\varepsilon^\gamma)}\mathbf{1}(B_y)\right]\\
&\quad\geqslant\mathbb{P}\left(\varepsilon W_1\notin D_0\right)\left(\mathbb{E}\left[e^{-\theta\lambda(\varepsilon)T_1}\right]-\mathbb{E}\left[e^{-\theta\lambda(\varepsilon)T_1}\mathbf{1}\{T_1<s_{r^*}+T_{rec}+\kappa\gamma|\ln\varepsilon|\}\right]\right)
\end{aligned}
$$

$$
\begin{aligned}
&-\mathbb{E}\left[e^{-\theta\lambda(\varepsilon)T_1} \sup_{y\in\tilde{D}(\varepsilon^\gamma)} \mathbf{1}(E^c(y))\right]\\
&\geqslant \frac{\lambda(\varepsilon)}{\beta_\varepsilon}\left(\frac{\beta_\varepsilon}{\theta\lambda(\varepsilon)+\beta_\varepsilon} - \frac{C}{5}\frac{\beta_\varepsilon}{\theta\lambda(\varepsilon)+\beta_\varepsilon}\frac{\lambda(\varepsilon)}{\beta_\varepsilon} - \frac{C}{5}\frac{\beta_\varepsilon}{\theta\lambda(\varepsilon)+\beta_\varepsilon}\frac{\lambda(\varepsilon)}{\beta_\varepsilon}\right)\\
&\geqslant \frac{\beta_\varepsilon}{\theta\lambda(\varepsilon)+\beta_\varepsilon}\left((1-C)\frac{\lambda(\varepsilon)}{\beta_\varepsilon}\right).
\end{aligned}
$$

Combining the estimates just obtained in Claim 1 and Claim 2 we finally get a lower estimate by a geometric series, leading to

$$
\begin{aligned}
&\mathbb{E}\left[\inf_{x\in\hat{D}(\varepsilon^\gamma)} e^{-\theta\lambda(\varepsilon)\tau_x(\varepsilon)}\right]\\
&\geqslant \sum_{k=1}^{\infty}\left(\frac{\beta_\varepsilon}{\beta_\varepsilon+\theta\lambda(\varepsilon)}\left(1-(1+C)\frac{\lambda(\varepsilon)}{\beta_\varepsilon}\right)\right)^{k-1}\frac{\beta_\varepsilon}{\theta\lambda(\varepsilon)+\beta_\varepsilon}\left((1-C)\frac{\lambda(\varepsilon)}{\beta_\varepsilon}\right)\\
&= \frac{\lambda(\varepsilon)(1-C)}{\theta\lambda(\varepsilon)-(1+C)\lambda(\varepsilon)} = \frac{1-C}{\theta+1+C}.
\end{aligned}
$$

The series converges if and only if $-(1+C) < \theta$. This completes the proof of our main Theorem. □

5.2.3 Asymptotic Exit Times in Probability

In this subsection we construct explicitly a family $(s^{\pm}(\varepsilon))_{\varepsilon>0}$ of random variables exponential law with parameter $\lambda^{\pm}(\varepsilon)$, to which the first exit times $\left(\tau_x^{\pm}(\varepsilon)\right)_{\varepsilon>0}$ converge in probability.

Definition 5.14. Recall that $W_k = \Delta_{T_k}\eta^\varepsilon, k \in \mathbb{N}$ is the k-th "large" jump of $(L(t))_{t\geqslant 0}$ in the sense of Sect. 2.1. For the event $B_k^\diamond(\varepsilon) = \{\varepsilon W_k \notin D_0^{\pm}\}, k \in \mathbb{N}$, $\varepsilon > 0$ we define the random variable

$$
s^{\pm}(\varepsilon) := \sum_{k=1}^{\infty} T_k \prod_{j=1}^{k-1}(1-\mathbf{1}(B_j^\diamond(\varepsilon)))\mathbf{1}(B_k^\diamond(\varepsilon)), \quad \varepsilon > 0. \tag{5.41}
$$

The distribution of $s^{\pm}(\varepsilon)$ can be computed directly.

Lemma 5.15. *For $\rho, \varepsilon \in (0,1)$ the random variable $s^{\pm}(\varepsilon)$ is exponentially distributed with parameter $\lambda^{\pm}(\varepsilon)$, $\lambda^{\pm}(\varepsilon) = \nu\left(\frac{1}{\varepsilon}D_0^{\pm}\right)$, where ν is the Lévy jump measure of the noise process $(L(t))_{t\geqslant 0}$ driving X^ε.*

Proof. Let $\theta > 0$. We can calculate the Laplace transform of $s^{\pm}(\varepsilon)$ directly

$$\begin{aligned}\mathbb{E}\left[e^{-\theta s^{\pm}(\varepsilon)}\right] &= \mathbb{E}\left[e^{-\theta \sum_{k=1}^{\infty} T_k \prod_{j=1}^{k-1}(1-\mathbf{1}(B_j^{\diamond}))\mathbf{1}(B_k^{\diamond})}\right] \\ &= \mathbb{E}\left[\prod_{k=1}^{\infty} e^{-\theta T_k \prod_{j=1}^{k-1}(1-\mathbf{1}(B_j^{\diamond}))\mathbf{1}(B_k^{\diamond})}\right] \\ &= \sum_{k=1}^{\infty} \mathbb{E}\left[e^{-\theta T_k} \prod_{j=1}^{k-1}(1-\mathbf{1}(B_j^{\diamond}))\mathbf{1}(B_k^{\diamond})\right] \\ &= \sum_{k=1}^{\infty} \mathbb{E}\left[\prod_{j=1}^{k-1} e^{-\theta t_j}(1-\mathbf{1}(B_j^{\diamond}))e^{-\theta t_k}\mathbf{1}(B_k^{\diamond})\right].\end{aligned}$$

Exploiting the independence of $(W_k)_{k\in\mathbb{N}}$ and $(T_k)_{k\in\mathbb{N}}$ as well as the stationarity of $(W_k)_{k\in\mathbb{N}}$ each summand takes the form

$$\begin{aligned}&\mathbb{E}\left[\prod_{j=1}^{k-1} e^{-\theta t_j}(1-\mathbf{1}(B_j^{\diamond}))e^{-\theta t_k}\mathbf{1}(B_k^{\diamond})\right] \\ &= \prod_{j=1}^{k-1} \mathbb{E}\left[e^{-\theta t_j}(1-\mathbf{1}(B_j^{\diamond}))\right]\mathbb{E}\left[e^{-\theta t_k}\mathbf{1}(B_k^{\diamond})\right] \\ &= \mathbb{E}\left[e^{-\theta t_1}(1-\mathbf{1}(B_1^{\diamond}))\right]^{k-1}\mathbb{E}\left[e^{-\theta t_1}\mathbf{1}(B_1^{\diamond})\right] \\ &= \left(\mathbb{E}\left[e^{-\theta t_1}\right](1-\mathbb{P}(B_1^{\diamond}))\right)^{k-1}\mathbb{E}\left[e^{-\theta t_1}\right]\mathbb{P}(B_1^{\diamond}) \\ &= \left(\frac{\beta_\varepsilon}{\theta+\beta_\varepsilon}(1-\frac{\lambda^{\pm}(\varepsilon)}{\beta_\varepsilon})\right)^{k-1}\frac{\beta_\varepsilon}{\theta+\beta_\varepsilon}\frac{\lambda^{\pm}(\varepsilon)}{\beta_\varepsilon}.\end{aligned}$$

Hence

$$\begin{aligned}\mathbb{E}\left[e^{-\theta\sigma^{\pm}(\varepsilon)}\right] &= \sum_{k=1}^{\infty}\left(\frac{\beta_\varepsilon}{\theta+\beta_\varepsilon}(1-\frac{\lambda^{\pm}(\varepsilon)}{\beta_\varepsilon})\right)^{k-1}\frac{\beta_\varepsilon}{\theta+\beta_\varepsilon}\frac{\lambda^{\pm}(\varepsilon)}{\beta_\varepsilon} \\ &= \frac{\beta_\varepsilon}{\theta+\beta_\varepsilon}\frac{\lambda^{\pm}(\varepsilon)}{\beta_\varepsilon}\frac{1}{1-\frac{\beta_\varepsilon}{\theta+\beta_\varepsilon}(1-\frac{\lambda^{\pm}(\varepsilon)}{\beta_\varepsilon})} = \frac{\lambda^{\pm}(\varepsilon)}{\beta_\varepsilon}\frac{1}{\frac{\theta+\beta_\varepsilon}{\beta_\varepsilon}-(1-\frac{\lambda^{\pm}(\varepsilon)}{\beta_\varepsilon})} \\ &= \frac{\lambda^{\pm}(\varepsilon)}{\theta+\lambda^{\pm}(\varepsilon)}.\end{aligned}$$

□

Theorem 5.16. *Suppose Hypotheses (H.1) and (H.2) and Convention (C) are satisfied. Then $s^{\pm}(\varepsilon)$ satisfies that for any $\theta > 0$ and $C > 0$ there is $\varepsilon_0 > 0$ such that for $0 < \varepsilon \leqslant \varepsilon_0$*

$$\mathbb{E}\left[\inf_{x \in \hat{D}^{\pm}(\varepsilon^{\gamma})} e^{-\theta|\tau_x^{\pm}(\varepsilon) - s^{\pm}(\varepsilon)|}\right] \geqslant 1 - C. \tag{5.42}$$

Corollary 5.17. *Under the assumptions of Theorem 5.16 there is a family of exponentially distributed random variables $(\bar{\tau}(\varepsilon))_{\varepsilon>0}$ with parameter 1 on the same probability space $(\Omega, \mathcal{F}, \mathbb{P})$ as the driving Lévy noise $(L(t))_{t\geqslant 0}$ such that in probability*

$$\lim_{\varepsilon \to 0+} \inf_{y \in \hat{D}^{\pm}(\varepsilon^{\gamma})} |\lambda^{\pm}(\varepsilon)\tau_y^{\pm}(\varepsilon) - \bar{\tau}(\varepsilon)| = \lim_{\varepsilon \to 0+} \sup_{y \in \hat{D}^{\pm}(\varepsilon^{\gamma})} |\lambda^{\pm}(\varepsilon)\tau_y^{\pm}(\varepsilon) - \bar{\tau}(\varepsilon)| = 0.$$

Proof. By Lemma 5.15 the random variable $s^{\pm}(\varepsilon), \varepsilon > 0$ defined by (5.41) is an exponentially distributed random variable with parameter $\lambda^{\pm}(\varepsilon)$. Hence $\bar{\tau}(\varepsilon) := \lambda^{\pm}(\varepsilon)s^{\pm}(\varepsilon)$, $\varepsilon > 0$, is exponentially distributed with parameter 1. Theorem 5.16 shows that for any $\theta > 0$ and $C > 0$ there is $\varepsilon_0 > 0$ such that for $0 < \varepsilon \leqslant \varepsilon_0$ and $\theta > 0$ we have

$$1 - C \leqslant \mathbb{E}\left[\inf_{y \in \hat{D}^{\pm}(\varepsilon^{\gamma})} e^{-\theta|\tau_y^{\pm}(\varepsilon) - \frac{\bar{\tau}(\varepsilon)}{\lambda^{\pm}(\varepsilon)}|}\right] \leqslant \mathbb{E}\left[\sup_{y \in \hat{D}^{\pm}(\varepsilon^{\gamma})} e^{-\theta|\tau_y^{\pm}(\varepsilon) - \frac{\bar{\tau}(\varepsilon)}{\lambda^{\pm}(\varepsilon)}|}\right] \leqslant 1. \tag{5.43}$$

Since $0 < \lambda^{\pm}(\varepsilon) = \varepsilon^{\alpha}\ell\left(\frac{1}{\varepsilon}\right)\mu\left((D_0^{\pm})^c\right) \searrow 0$ for $\varepsilon \to 0+$, we may fix $\varepsilon_0 > 0$ such that $\lambda^{\pm}(\varepsilon) < 1$ for $0 < \varepsilon \leqslant \varepsilon_0$. By monotonicity we obtain for this $\varepsilon_0 > 0$ and the same $C > 0$, $\theta > 0$ as before that

$$1 - C \leqslant \mathbb{E}\left[\inf_{y \in \hat{D}^{\pm}(\varepsilon^{\gamma})} e^{-\theta|\lambda^{\pm}(\varepsilon)\tau_y^{\pm}(\varepsilon) - \bar{\tau}(\varepsilon)|}\right] \leqslant \mathbb{E}\left[\sup_{y \in \hat{D}^{\pm}(\varepsilon^{\gamma})} e^{-\theta|\lambda^{\pm}(\varepsilon)\tau_y^{\pm}(\varepsilon) - \bar{\tau}(\varepsilon)|}\right] \leqslant 1 \tag{5.44}$$

for $0 < \varepsilon \leqslant \varepsilon_0$. Hence

$$\lim_{\varepsilon \to 0+} \mathcal{L}\Big(\inf_{y \in \hat{D}^{\pm}(\varepsilon^{\gamma})} |\lambda^{\pm}(\varepsilon)\tau_y^{\pm}(\varepsilon) - \bar{\tau}(\varepsilon)|\Big) = \lim_{\varepsilon \to 0+} \mathcal{L}\Big(\sup_{y \in \hat{D}^{\pm}(\varepsilon^{\gamma})} |\lambda^{\pm}(\varepsilon)\tau_y^{\pm}(\varepsilon) - \bar{\tau}(\varepsilon)|\Big) = \delta_0$$

in the weak sense. Due to the equivalence of weak convergence and convergence in probability for constant limits we obtain the desired result. □

Proof (of Theorem 5.16).
Step 1: Reduction to incremental events
Recall for $\varepsilon > 0$ the definition of the events

$$A_j^{\diamond} = \left(B_j^{\diamond}\right)^c = \{\varepsilon W_j \in D_0^{\pm}\}, \text{ for } j \in \mathbb{N}.$$

In this step we follow the lines of the first part of the proof of Theorem 5.12. Exploiting the strong Markov property of X^ε and Y^ε respectively and the Lévy property of L we may estimate the exit first exit time from below by probabilities of exit events A_x^-, B_x, $A_1^\diamond$ and $B_1^\diamond$ in the following way

$$
\begin{aligned}
&\mathbb{E}\Big[\inf_{x\in\hat{D}^\pm(\varepsilon^\gamma)} e^{-\theta|\tau_x^\pm(\varepsilon)-s^\pm(\varepsilon)|}\Big] \\
&\geqslant \sum_{k=1}^\infty \mathbb{E}\Big[\inf_{x\in\hat{D}^\pm(\varepsilon^\gamma)} e^{-\theta|\tau_x^\pm(\varepsilon)-s^\pm(\varepsilon)|}\mathbf{1}(\{\tau_x^\pm = T_k\}\cap\bigcap_{j=1}^{k-1} A_{j-1}^\diamond \cap B_k^\diamond)\Big] \\
&= \sum_{k=1}^\infty \mathbb{E}\Big[\inf_{x\in\hat{D}^\pm(\varepsilon^\gamma)} \mathbf{1}\{\tau_x^\pm = T_k\}\cap\bigcap_{j=1}^{k-1} A_{j-1}^\diamond \cap B_k^\diamond\Big] \\
&\geqslant \sum_{k=1}^\infty \mathbb{E}\Big[\inf_{x\in\hat{D}^\pm(\varepsilon^\gamma)} \mathbf{1}(\bigcap_{j=1}^{k-1} A^-_{X^\varepsilon(0;x)}\circ\theta_{T_{j-1}} \cap B_{X^\varepsilon(0;x)}\circ\theta_{T_{k-1}} \cap \bigcap_{j=1}^{k-1} A_{j-1}^\diamond\cap B_k^\diamond)\Big] \\
&\geqslant \sum_{k=1}^\infty \mathbb{E}\Big[\inf_{x\in\hat{D}^\pm(\varepsilon^\gamma)} \prod_{j=1}^{k-1}\Big(\mathbf{1}(A^-_{X^\varepsilon(0;x)}\circ\theta_{T_{j-1}}\cap A_j^\diamond)\Big)\mathbf{1}(B_{X^\varepsilon(0;x)}\circ\theta_{T_{k-1}}\cap B_k^\diamond)\Big].
\end{aligned}
$$

With the help of the strong Markov property of X^ε we obtain for $k\in\mathbb{N}$ that

$$
\begin{aligned}
&\mathbb{E}\Big[\inf_{x\in\hat{D}^\pm(\varepsilon^\gamma)} \prod_{j=1}^{k-1}\Big(\mathbf{1}(A^-_{X^\varepsilon(0;x)}\circ\theta_{T_{j-1}}\cap A_j^\diamond)\Big)\mathbf{1}(B_{X^\varepsilon(0;x)}\circ\theta_{T_{k-1}}\cap B_k^\diamond)\Big] \\
&\geqslant \mathbb{E}\Big[\mathbb{E}\Big[\inf_{x\in\hat{D}^\pm(\varepsilon^\gamma)} \prod_{j=1}^{k-1}\Big(\mathbf{1}(A^-_{X^\varepsilon(0;x)}\circ\theta_{T_{j-1}}\cap A_j^\diamond)\Big)\mathbf{1}(B_{X^\varepsilon(0;x)}\circ\theta_{T_{k-1}}\cap B_k^\diamond)|\,\mathscr{F}_{T_{k-1}}\Big]\Big] \\
&\geqslant \mathbb{E}\Big[\inf_{x\in\hat{D}^\pm(\varepsilon^\gamma)} \prod_{j=1}^{k-2}\Big(\mathbf{1}(A^-_{X^\varepsilon(0;x)}\circ\theta_{T_{j-1}}\cap A_j^\diamond)\Big)\mathbf{1}(A^-_{X^\varepsilon(0;x)}\circ\theta_{T_{k-2}}\cap A_{k-1}^\diamond)\Big] \\
&\quad \mathbb{E}\Big[\inf_{y\in\hat{D}^\pm(\varepsilon^\gamma)} \mathbf{1}(B_y\cap B_1^\diamond)\Big].
\end{aligned}
$$

By $k-1$-fold iteration of this argument we obtain for $k\in\mathbb{N}$

$$
\begin{aligned}
&\mathbb{E}\Big[\inf_{x\in\hat{D}^\pm(\varepsilon^\gamma)} \prod_{j=1}^{k-1}\Big(\mathbf{1}(A^-_{X^\varepsilon(0;x)}\circ\theta_{T_{j-1}}\cap A_j^\diamond)\Big)\mathbf{1}(B_{X^\varepsilon(0;x)}\circ\theta_{T_{k-1}}\cap B_k^\diamond)\Big] \\
&\quad\geqslant \mathbb{E}\Big[\inf_{y\in\hat{D}^\pm(\varepsilon^\gamma)} \mathbf{1}(A_y^-\cap A_1^\diamond)\Big]^{k-1}\mathbb{E}\Big[\inf_{y\in\hat{D}^\pm(\varepsilon^\gamma)} \mathbf{1}(B_y\cap B_1^\diamond)\Big]. \qquad (5.45)
\end{aligned}
$$

Step 2: Inspection of the incremental events
This step consists in the estimate of the events $A_y^- \cap A_1^\diamond$ and $B_y \cap B_1^\diamond$ by the small deviation event E_x, the relaxation event $\{T_1 \geqslant s_{r*} + T_{rec} + \kappa\gamma|\ln\varepsilon|\}$ and the pure jump events $\{\varepsilon W_1 \in D_0^\pm(\varepsilon^\gamma, \varepsilon^{2\gamma}, \varepsilon^{2\gamma}, \varepsilon^{2\gamma})\}$ and $A_1^\diamond$. By Lemma 5.2 (*xv*) and (*xvi*) we know for $x \in \hat{D}^\pm(\varepsilon^\gamma)$

$$\mathbf{1}(A_x^- \cap A_1^\diamond) \geqslant \mathbf{1}\{\varepsilon W_1 \in D_0^\pm(\varepsilon^\gamma, \varepsilon^{2\gamma}, \varepsilon^{2\gamma}, \varepsilon^{2\gamma})\} - \mathbf{1}\{T_1 < s_{r*} + T_{rec} + \kappa\gamma|\ln\varepsilon|\} - \mathbf{1}(E_x^c), \tag{5.46}$$

$$\mathbf{1}(B_x \cap B_1^\diamond) \geqslant \mathbf{1}\{\varepsilon W_1 \notin D_0^\pm\}(1 - \mathbf{1}\{T_1 < s_{r*} + T_{rec} + \kappa\gamma|\ln\varepsilon|\}) - \mathbf{1}(E_x^c). \tag{5.47}$$

Step 3: Lower estimate of the first factor
In this step we exploit Lemma 5.5 (V) and Corollary 4.6 in order to estimate the first factor in the summands of the right-hand side of (5.28)

$$\begin{aligned}
&\mathbb{E}\left[\inf_{y \in \hat{D}^\pm(\varepsilon^\gamma)} \mathbf{1}_{A_y^- \cap A_1^\diamond}\right] \\
&\quad \geqslant \mathbb{P}\left(\varepsilon W_1 \in D_0^\pm(\varepsilon^\gamma, \varepsilon^{2\gamma}, \varepsilon^{2\gamma}, \varepsilon^{2\gamma})\right) - \mathbb{P}\left(T_1 < s_{r*} + T_{rec} + \kappa\gamma|\ln\varepsilon|\right) \\
&\quad - \mathbb{E}\left[\inf_{y \in \tilde{D}^\pm(\varepsilon^\gamma)} \mathbf{1}(E_y^c)\right].
\end{aligned}$$

By Lemma 5.5 (V) for any $C_1 > 0$ given there is $\varepsilon_1 > 0$ such that such that for $0 < \varepsilon \leqslant \varepsilon_1$

$$\mathbb{P}(\varepsilon W_1 \in D_0^c(\varepsilon^\gamma, \varepsilon^{2\gamma}, \varepsilon^{2\gamma}, \varepsilon^{2\gamma})) \leqslant (1 + C_1)\frac{\lambda^\pm(\varepsilon)}{\beta_\varepsilon}.$$

Furthermore for any $C_2 > 0$ there is $\varepsilon_2 > 0$ such that for $0 < \varepsilon \leqslant \varepsilon_2$

$$\begin{aligned}
\mathbb{P}\left(T_1 < s_{r*} + T_{rec} + \kappa\gamma|\ln\varepsilon|\right) &= \int_0^{s_{r*} + T_{rec} + \kappa\gamma|\ln\varepsilon|} \beta_\varepsilon e^{-\beta_\varepsilon s} ds \\
&= 1 - e^{-\beta_\varepsilon(s_{r*} + T_{rec} + \kappa\gamma|\ln\varepsilon|)} \\
&\leqslant (1 + C_2)\beta_\varepsilon(s_{r*} + T_{rec} + \kappa\gamma|\ln\varepsilon|).
\end{aligned}$$

In addition by Corollary 4.6 for any $C_3 > 0$ there is $\varepsilon_3 > 0$ such that for $0 < \varepsilon \leqslant \varepsilon_3$

$$\mathbb{E}\left[\inf_{y \in \tilde{D}^\pm(\varepsilon^\gamma)} \mathbf{1}(E_y^c)\right] \leqslant C_3 \frac{\lambda^\pm(\varepsilon)}{\beta_\varepsilon}.$$

This means if $C_i \leqslant \frac{J_1}{4}$, $i = 1, 2, 3$ that for $0 < \varepsilon \leqslant \min\{\varepsilon_1, \varepsilon_2, \varepsilon_3\}$ respectively

$$\begin{aligned}
&\mathbb{E}\left[\inf_{y \in \hat{D}^{\pm}(\varepsilon^{\gamma})} \mathbf{1}(A_y^- \cap A_1^{\diamond})\right] \\
&\geqslant 1 - (1 + C_1)\frac{\lambda^{\pm}(\varepsilon)}{\beta_\varepsilon} - (1 + C_2)\beta_\varepsilon(s_{r^*} + T_{rec} + \kappa\gamma|\ln\varepsilon|) - 2C_3\frac{\lambda^{\pm}(\varepsilon)}{\beta_\varepsilon} \\
&\geqslant 1 - (1 + J_1)\frac{\lambda^{\pm}(\varepsilon)}{\beta_\varepsilon}. \qquad (5.48)
\end{aligned}$$

Step 4: Lower estimate of the second factor

Since $\rho > \frac{1}{2}$, we know $\frac{\beta_\varepsilon}{\lambda^{\pm}(\varepsilon)}\beta_\varepsilon(s_{r^*} + T_{rec} + \kappa\gamma|\ln\varepsilon|) \to 0+$, $\varepsilon \to 0+$. Hence for given $C_4 > 0$ there is $\varepsilon_4 > 0$ with $\frac{\beta_\varepsilon}{\lambda^{\pm}(\varepsilon)}\beta_\varepsilon(s_{r^*} + T_{rec} + \kappa\gamma|\ln\varepsilon|) \leqslant C_4$ for $0 < \varepsilon \leqslant \varepsilon_4$. Thus for $C_i \leqslant \frac{J_2}{3}$, $i = 3, 4$ and $0 < \varepsilon \leqslant \min\{\varepsilon_4, \varepsilon_5\}$

$$\begin{aligned}
&\mathbb{E}\left[\inf_{y \in \hat{D}^{\pm}(\varepsilon^{\gamma})} \mathbf{1}(B_y \cap B_1^{\diamond})\right] \\
&\geqslant \mathbb{P}(\varepsilon W_1 \notin D_0^{\pm}) - \mathbb{P}(T_1 < s_{r^*} + T_{rec} + \kappa\gamma|\ln\varepsilon|) - \mathbb{E}\left[\inf_{y \in \tilde{D}^{\pm}(\varepsilon^{\gamma})} \mathbf{1}(E_y^c)\right] \\
&\geqslant \frac{\lambda^{\pm}(\varepsilon)}{\beta_\varepsilon} - (1 + C_4)\beta_\varepsilon(s_{r^*} + T_{rec} + \kappa\gamma|\ln\varepsilon|) - C_3\frac{\lambda^{\pm}(\varepsilon)}{\beta_\varepsilon} \\
&\geqslant (1 - J_2)\frac{\lambda^{\pm}(\varepsilon)}{\beta_\varepsilon}.
\end{aligned}$$

Step 5: The asymptotic geometric series

We now combine Steps 4 and 5 with estimate (5.28) for $0 < \varepsilon \leqslant \min\{\varepsilon_1, \varepsilon_2, \varepsilon_3, \varepsilon_4\}$, such that

$$\sum_{k=1}^{\infty}\left(1 - (1 + J_1)\frac{\lambda^{\pm}(\varepsilon)}{\beta_\varepsilon}\right)^{k-1}(1 - J_2)\frac{\lambda^{\pm}(\varepsilon)}{\beta_\varepsilon} \geqslant \frac{1 - J_2}{1 + J_1} \geqslant 1 - C$$

if $J_1 > 0$ and $J_2 > 0$ are chosen to satisfy $0 \leqslant J_2 \leqslant C + (1 - C)J_1$. □

Chapter 6
Asymptotic Transition Times

The preceding chapter is concerned with the effect of small Lévy noise in H of intensity ε of triggering exits from the reduced domains of attraction of the stable states $\phi^{\pm}$ of a Chafee–Infante equation. Noise is seen quite generally to make stable states of deterministic systems given by ordinary or partial differential equations metastable. In this chapter, we shall investigate more closely the dynamics of the stochastic system, in particular the stochastic transition and wandering behavior between the metastable states. We shall ask questions about the reduced dynamics of the system, i.e. the reduction of the jump diffusion equation to a simple Markov chain in the small noise limit $\varepsilon \to 0+$ boiling down the dynamics to a simple switching between the metastable states. It will be seen that this reduction is related to a scaling limit of the jump diffusion in the polynomial scale $\varepsilon^{-\alpha}$ resulting from the asymptotic behavior of first exit times of domains of attraction encountered in the previous chapter.

6.1 Asymptotic Times to Enter Different Reduced Domains of Attraction

For $\varepsilon, \gamma > 0$ we recall the complement of the reduced domains of attraction

$$\hat{D}^0(\varepsilon^\gamma) = H \setminus \left(\hat{D}^+(\varepsilon^\gamma) \cup \hat{D}^-(\varepsilon^\gamma)\right),$$

with $\hat{D}^{\pm}(\varepsilon^\gamma) = D^{\pm}(\varepsilon^\gamma, \varepsilon^{2\gamma}, \varepsilon^{2\gamma})$ according to Definition 2.9. Theorems 5.8 and 5.11 describe the asymptotic behavior of the first exit times $\tau_x^{\pm}(\varepsilon)$ from $\tilde{D}^{\pm}(\varepsilon^\gamma)$ for initial values in $x \in \hat{D}^{\pm}(\varepsilon^\gamma)$. The aim of this and the next section is to determine the asymptotic behavior of the transition times between small balls centered in the metastable states. In a first step we consider first exit times from

$$\hat{D}^{\pm 0}(\varepsilon^\gamma) := \hat{D}^{\pm}(\varepsilon^\gamma) \cup \hat{D}^0(\varepsilon^\gamma) = H \setminus (\hat{D}^+(\varepsilon^\gamma) \cup \hat{D}^+(\varepsilon^\gamma)).$$

A. Debussche et al., *The Dynamics of Nonlinear Reaction-Diffusion Equations with Small Lévy Noise*, Lecture Notes in Mathematics 2085, DOI 10.1007/978-3-319-00828-8_6,

Definition 6.1. Define for $x \in \hat{D}^{\pm}(\varepsilon^{\gamma})$

$$\tau_x^{\pm 0}(\varepsilon) := \inf\{t > 0 \mid X^{\varepsilon}(t;x) \notin \hat{D}^{\pm 0}(\varepsilon^{\gamma})\}$$

the first exit time of $X^{\varepsilon}(\cdot;x)$ to leave the enhanced domain of attraction $\hat{D}^{\pm 0}(\varepsilon^{\gamma})$.

We shall show that the slow deterministic dynamics close to the separatrix $\mathscr{S}$ asymptotically has no contribution to the exit rate, that is $\tau^{\pm}(\varepsilon) \approx \tau^{\pm 0}(\varepsilon)$ for $\varepsilon \to 0+$.

Analogously to (5.2) in Chap. 5 we define for $y \in \tilde{D}^{\pm}(\varepsilon^{\gamma})$ the modified exit event

$$\hat{B}_y := \left\{\omega \in \Omega \mid Y^{\varepsilon}(s;y) \in \tilde{D}^{\pm}(\varepsilon^{\gamma}) \text{ for } s \in [0,T_1] \text{ and } Y^{\varepsilon}(T_1;y) + \varepsilon W_1 \notin \hat{D}^{\pm 0}(\varepsilon^{\gamma})\right\}. \tag{6.1}$$

We need the following slight modification of Lemma 5.2.

Lemma 6.2. *For $\rho \in (1/2, 1), \gamma \in (0, 1-\rho)$ there exists $\varepsilon_0 > 0$ such that for all $0 < \varepsilon \leqslant \varepsilon_0, \kappa > 0,\ y \in \hat{D}^{\pm}(\varepsilon^{\gamma})$*

$$\begin{aligned}\mathbf{1}(\hat{B}_y) \geqslant\ & \mathbf{1}\{\varepsilon W_1 \in D^{\mp}(\varepsilon^{\gamma}, \varepsilon^{2\gamma}, \varepsilon^{2\gamma}, \varepsilon^{2\gamma}) - \phi^{\pm}\}(1 - \mathbf{1}\{T_1 < s_{r^*} + T_{rec} + \kappa\gamma|\ln\varepsilon|\}) \\ & - \mathbf{1}(E_y^c).\end{aligned} \tag{6.2}$$

Proof. In order to estimate $\mathbf{1}(\hat{B}_y)$ for $y \in \hat{D}(\varepsilon^{\gamma})$ we need the following estimates analogous to statement $vi)$ of Lemma 5.1, namely

$$\mathbf{1}(E_y)\mathbf{1}\{T_1 \geqslant s_{r^*} + T_{rec} + \kappa\gamma|\ln\varepsilon|\}\mathbf{1}\{\varepsilon W_1 \in D^{\mp}(\varepsilon^{\gamma}, \varepsilon^{2\gamma}, \varepsilon^{2\gamma}, \varepsilon^{2\gamma}) - \phi^{\pm}\} \leqslant \mathbf{1}(\hat{B}_y). \tag{6.3}$$

Proposition 2.12 states that on the event $\{T_1 \geqslant s_{r^*} + T_{rec} + \kappa\gamma|\ln\varepsilon|\}$ we have $u(T_1;y) \in B_{(1/2)\varepsilon^{2\gamma}}(\phi^{\pm})$. Hence on $\{T_1 \geqslant s_{r^*} + T_{rec} + \kappa\gamma|\ln\varepsilon|\} \cap E_y$ the relationship $Y^{\varepsilon}(T_1;y) \in B_{\varepsilon^{2\gamma}}(\phi^{\pm})$ holds, exactly as in step 1 of the proof of Lemma 5.1. Hence in order to arrive in $\hat{B}_y$ we only have to ensure that

$$B_{\varepsilon^{2\gamma}}(\phi^{\pm}) + D^{\mp}(\varepsilon^{\gamma}, \varepsilon^{2\gamma}, \varepsilon^{2\gamma}, \varepsilon^{2\gamma}) - \phi^{+} \subseteq \hat{D}^{\mp}(\varepsilon^{\gamma}).$$

The shifts by $\phi^{\pm}$ cancel out, reducing the inclusion to

$$B_{\varepsilon^{2\gamma}}(0) + D^{\mp}(\varepsilon^{\gamma}, \varepsilon^{2\gamma}, \varepsilon^{2\gamma}, \varepsilon^{2\gamma}) \subseteq \hat{D}^{\mp}(\varepsilon^{\gamma}),$$

which is a result of Lemma 2.11. This proves the estimate (6.3). By (6.3), the estimate (6.2) follows from the calculation

$$\begin{aligned}\mathbf{1}(\hat{B}_y) &\geqslant \mathbf{1}(\hat{B}_y)\mathbf{1}(E_y)\mathbf{1}\{T_1 \geqslant s_{r^*} + T_{rec} + \kappa\gamma|\ln\varepsilon|\} \\ &\geqslant \mathbf{1}\{\varepsilon W_1 \in D^{\mp}(\varepsilon^{\gamma}, \varepsilon^{2\gamma}, \varepsilon^{2\gamma}, \varepsilon^{2\gamma}) - \phi^{\pm}\}\mathbf{1}(E_y)\mathbf{1}\{T_1 \geqslant s_{r^*} + T_{rec} + \kappa\gamma|\ln\varepsilon|\}\end{aligned}$$

$$\geqslant \mathbf{1}\{\varepsilon W_1 \in D^{\mp}(\varepsilon^{\gamma}, \varepsilon^{2\gamma}, \varepsilon^{2\gamma}, \varepsilon^{2\gamma}) - \phi^{\pm}\}$$
$$(1 - \mathbf{1}\{T_1 < s_{r^*} + T_{rec} + \kappa\gamma|\ln \varepsilon|\}) - \mathbf{1}(E_y^c). \qquad \square$$

Theorem 6.3. *Suppose Hypotheses (H.1) and (H.2) and Convention (C) to be satisfied. Then the family of random variables $(s^{\pm}(\varepsilon))_{\varepsilon>0}$ defined by (5.41) satisfies that for any $\theta > 0$ and $0 < C < 1$ there is $\varepsilon_0 > 0$ such that for $0 < \varepsilon \leqslant \varepsilon_0$*

$$\mathbb{E}\left[\inf_{x\in\hat{D}^{\pm}(\varepsilon^{\gamma})} e^{-\theta|\tau_x^{\pm 0}(\varepsilon)-s^{\pm}(\varepsilon)|}\right] \geqslant 1 - C. \tag{6.4}$$

Proof (of Theorem 6.3).
Step 1: Reduction to incremental events
Recall for $\varepsilon > 0$ the events

$$A_j^{\diamond} := \left(B_j^{\diamond}\right)^c = \{\varepsilon W_j \in D_0^{\pm}\}, \text{ for } j \in \mathbb{N}.$$

In this step we follow the lines of the first part proof of Theorem 5.16. Exploiting the identity

$$\{X^{\varepsilon}(\tau_x^{\pm}(\varepsilon); x) \notin \hat{D}^{\pm 0}(\varepsilon^{\gamma})\} = \{\tau_x^{\pm}(\varepsilon) = \tau_x^{\pm 0}\},$$

the strong Markov property of X^{ε} and Y^{ε} respectively and the Lévy property of L we may estimate the first exit time from below by probabilities of exit events A_x^-, $\hat{B}_x$, $A_1^{\diamond}$ and $B_1^{\diamond}$ such that

$$\begin{aligned}
&\mathbb{E}\left[\inf_{x\in\hat{D}^{\pm}(\varepsilon^{\gamma})} e^{-\theta|\tau_x^{\pm 0}(\varepsilon)-s^{\pm}(\varepsilon)|}\right] \\
&\quad\geqslant \mathbb{E}\left[\inf_{x\in\hat{D}^{\pm}(\varepsilon^{\gamma})} e^{-\theta|\tau_x^{\pm 0}(\varepsilon)-s^{\pm}(\varepsilon)|}\mathbf{1}\{X^{\varepsilon}(\tau_x^{\pm}(\varepsilon); x) \notin \hat{D}^{\pm 0}(\varepsilon^{\gamma})\}\right] \\
&\quad= \mathbb{E}\left[\inf_{x\in\hat{D}^{\pm}(\varepsilon^{\gamma})} e^{-\theta|\tau_x^{\pm}(\varepsilon)-s^{\pm}(\varepsilon)|}\mathbf{1}\{X^{\varepsilon}(\tau_x^{\pm}(\varepsilon); x) \notin \hat{D}^{\pm 0}(\varepsilon^{\gamma})\}\right] \\
&\quad= \sum_{k=1}^{\infty}\mathbb{E}\left[\inf_{x\in\hat{D}^{\pm}(\varepsilon^{\gamma})} \mathbf{1}_{\{\tau_x^{\pm}(\varepsilon)=T_k\}\cap\bigcap_{j=1}^{k-1}A_{j-1}^{\diamond}\cap B_k^{\diamond}}\mathbf{1}\{X^{\varepsilon}(\tau_x^{\pm}(\varepsilon); x) \notin \hat{D}^{\pm 0}(\varepsilon^{\gamma})\}\right] \\
&\quad\geqslant \sum_{k=1}^{\infty}\mathbb{E}\left[\inf_{x\in\hat{D}^{\pm}(\varepsilon^{\gamma})} \prod_{j=1}^{k-1}\left(\mathbf{1}_{A^-_{X^{\varepsilon}(0;x)}\circ\theta_{T_{j-1}}\cap A_j^{\diamond}}\right)\mathbf{1}_{\hat{B}_{X^{\varepsilon}(0;x)}\circ\theta_{T_{k-1}}\cap B_k^{\diamond}}\right].
\end{aligned}$$

By $k-1$-fold iterated application of the strong Markov property of X^ε as for example in the proof of Theorem 5.16 we obtain for $k \in \mathbb{N}$ that

$$\mathbb{E}\left[\inf_{x\in\hat{D}^{\pm}(\varepsilon^\gamma)} \prod_{j=1}^{k-1}\left(\mathbf{1}_{A^-_{X^\varepsilon(0;x)}\circ\theta_{T_{j-1}}\cap A^\diamond_j}\right)\mathbf{1}_{\hat{B}_{X^\varepsilon(0;x)}\circ\theta_{T_{k-1}}\cap B^\diamond_k}\right]$$
$$\geqslant \mathbb{E}\left[\inf_{y\in\hat{D}^{\pm}(\varepsilon^\gamma)}\mathbf{1}_{A^-_y\cap A^\diamond_1}\right]^{k-1}\mathbb{E}\left[\inf_{y\in\hat{D}^{\pm}(\varepsilon^\gamma)}\mathbf{1}_{\hat{B}_y\cap B^\diamond_1}\right]. \tag{6.5}$$

The lower estimate of the first factor is given by the asymptotic estimate (5.48) in Step 3 of the proof for Theorem 5.16, for which we keep the notation. Hence it only remains to estimate the second factor.

Step 2: Lower estimate of the second factor

By Lemma 6.2 estimate (6.2) we know for $y \in \hat{D}^{\pm}(\varepsilon^\gamma)$

$$\mathbf{1}(\hat{B}_y) \geqslant \mathbf{1}\{\varepsilon W_1 \in D^{\mp}(\varepsilon^\gamma,\varepsilon^{2\gamma},\varepsilon^{2\gamma},\varepsilon^{2\gamma}) - \phi^{\pm}\}$$
$$(1-\mathbf{1}\{T_1 < s_{r^*} + T_{rec} + \kappa\gamma|\ln\varepsilon|\}) - \mathbf{1}(E^c_y).$$

Since $\{\varepsilon W_1 \in D^{\mp}(\varepsilon^\gamma,\varepsilon^{2\gamma},\varepsilon^{2\gamma},\varepsilon^{2\gamma}) - \phi^{\pm}\} \subseteq \{\varepsilon W_1 \in (D^{\pm}_0)^c\} = B^\diamond$ we may infer that

$$\mathbf{1}(B^\diamond \cap \hat{B}_y)$$
$$\geqslant \mathbf{1}\{\varepsilon W_1 \in D^{\mp}(\varepsilon^\gamma,\varepsilon^{2\gamma},\varepsilon^{2\gamma},\varepsilon^{2\gamma}) - \phi^{\pm}\}(1-\mathbf{1}\{T_1 < s_{r^*} + T_{rec} + \kappa\gamma|\ln\varepsilon|\})$$
$$-\mathbf{1}(E^c_y \cap B^\diamond)$$
$$\geqslant \mathbf{1}\{\varepsilon W_1 \in D^{\mp}(\varepsilon^\gamma,\varepsilon^{2\gamma},\varepsilon^{2\gamma},\varepsilon^{2\gamma}) - \phi^{\pm}\}\mathbf{1}\{T_1 \geqslant s_{r^*} + T_{rec} + \kappa\gamma|\ln\varepsilon|\} - \mathbf{1}(E^c_y).$$

Passing to the infimum over $y \in \hat{D}^{\pm}(\varepsilon^\gamma)$ on the left-hand side, and correspondingly to the supremum on the right hand side, and then integrating yields

$$\mathbb{E}\left[\inf_{y\in\hat{D}^{\pm}(\varepsilon^\gamma)}\mathbf{1}(\hat{B}_y\cap B^\diamond)\right]$$
$$\geqslant \mathbb{P}\left(\varepsilon W_1 \in D^{\mp}(\varepsilon^\gamma,\varepsilon^{2\gamma},\varepsilon^{2\gamma},\varepsilon^{2\gamma}) - \phi^{\pm}\right)\,\mathbb{P}\left(T_1 \geqslant s_{r^*} + T_{rec} + \kappa\gamma|\ln\varepsilon|\}\right)$$
$$-\mathbb{E}\left[\sup_{y\in\tilde{D}(\varepsilon^\gamma)}\mathbf{1}(E^c_y)\right].$$

By the estimates in the proof of Propositions 5.16, 5.12, and 5.13 the asymptotic behavior of all terms arising in the preceding inequality with exception of

$$\mathbb{P}\left(\varepsilon W_1 \in D^{\mp}(\varepsilon^{\gamma}, \varepsilon^{2\gamma}, \varepsilon^{2\gamma}, \varepsilon^{2\gamma}) - \phi^{\pm}\right)$$

are well known. To estimate the latter, we exploit the representation

$$D^{\mp}(\varepsilon^{\gamma}, \varepsilon^{2\gamma}, \varepsilon^{2\gamma}, \varepsilon^{2\gamma}) - \phi^{\pm} = \left(D_0^{\pm}\right)^c \setminus \left(\left(D_0^{\pm}\right)^c \setminus \left(D^{\mp}(\varepsilon^{\gamma}, \varepsilon^{2\gamma}, \varepsilon^{2\gamma}, \varepsilon^{2\gamma}) - \phi^{\pm}\right)\right).$$

Note that by the regular variation of ν and Lemma 5.4 (iv) for each $0 < C_5 < 1$ there is $\varepsilon_5 > 0$ small enough, such that for $0 < \varepsilon \leqslant \varepsilon_5$

$$\begin{aligned}
&\mathbb{P}\left(\varepsilon W_1 \in D^{\mp}(\varepsilon^{\gamma}, \varepsilon^{2\gamma}, \varepsilon^{2\gamma}, \varepsilon^{2\gamma}) - \phi^{\pm}\right) \\
&= \frac{\lambda^{\pm}(\varepsilon)}{\beta_\varepsilon} - \frac{1}{\beta_\varepsilon}\nu\left(\frac{1}{\varepsilon}\left((D_0^{\pm})^c \setminus (D^{\mp}(\varepsilon^{\gamma}, \varepsilon^{2\gamma}, \varepsilon^{2\gamma}, \varepsilon^{2\gamma}) - \phi^{\pm})\right)\right) \\
&= \frac{\lambda^{\pm}(\varepsilon)}{\beta_\varepsilon} - \frac{\lambda^{\pm}(\varepsilon)}{\beta_\varepsilon}\frac{\nu\left(\frac{1}{\varepsilon}\left(\left((D^{\pm})^c \setminus D^{\mp}(\varepsilon^{\gamma}, \varepsilon^{2\gamma}, \varepsilon^{2\gamma}, \varepsilon^{2\gamma})\right) - \phi^{\pm}\right)\right)}{\lambda^{\pm}(\varepsilon)} \\
&\geqslant \frac{\lambda^{\pm}(\varepsilon)}{\beta_\varepsilon}\left(1 - \frac{\mu\left(\left((D^{\pm})^c \setminus D^{\mp}(\varepsilon_0^{\gamma}, \varepsilon_0^{2\gamma}, \varepsilon_0^{2\gamma}, \varepsilon_0^{2\gamma})\right) - \phi^{\pm}\right)}{\mu\left(\left(D_0^{\pm}\right)^c\right)}\right) \geqslant \frac{\lambda^{\pm}(\varepsilon)}{\beta_\varepsilon}(1 - C_5).
\end{aligned}$$

Thus keeping the notation from the proof of Theorem 5.16 we obtain for $C_i \leqslant \frac{J_2}{3} \leqslant \frac{1}{3}$, $i = 3, 4, 5$ and $0 < \varepsilon \leqslant \min\{\varepsilon_3, \varepsilon_4, \varepsilon_5\}$

$$\begin{aligned}
&\mathbb{E}\left[\inf_{y \in \tilde{D}^{\pm}(\varepsilon^{\gamma})} \mathbf{1}(\hat{B}_y \cap B^{\diamond})\right] \\
&\geqslant \mathbb{P}\left(\varepsilon W_1 \in D^{\mp}(\varepsilon^{\gamma}, \varepsilon^{2\gamma}, \varepsilon^{2\gamma}, \varepsilon^{2\gamma}) - \phi^{\pm}\right) \mathbb{P}\left(T_1 \geqslant s_{r^*} + T_{rec} + \kappa\gamma|\ln\varepsilon|\}\right) \\
&\quad - \mathbb{E}\left[\sup_{y \in \tilde{D}(\varepsilon^{\gamma})} \mathbf{1}(E_y^c)\right] \\
&\geqslant \frac{\lambda^{\pm}(\varepsilon)}{\beta_\varepsilon}(1 - C_5)(1 - C_4) - C_3\varepsilon^{\vartheta} \\
&\geqslant \frac{\lambda^{\pm}(\varepsilon)}{\beta_\varepsilon}(1 - J_2).
\end{aligned}$$

Step 3: The asymptotic geometric series
We can now estimate (6.5) for $0 < \varepsilon \leqslant \min\{\varepsilon_1, \varepsilon_2, \varepsilon_3, \varepsilon_4, \varepsilon_5\}$, such that

$$\sum_{k=1}^{\infty} \left(1 - (1 + J_1)\frac{\lambda^{\pm}(\varepsilon)}{\beta_{\varepsilon}}\right)^{k-1} (1 - J_2)\frac{\lambda^{\pm}(\varepsilon)}{\beta_{\varepsilon}} \geqslant \frac{1 - J_2}{1 + J_1} \geqslant 1 - C$$

if $J_1 > 0$ and $J_2 > 0$ are chosen to satisfy $0 \leqslant J_2 \leqslant C + (1 - C)J_1$. □

With an analogous proof as for Corollary 5.17 we obtain the following statement.

Corollary 6.4. *Under the assumptions of Theorem 6.3 there is a family random variables $(\bar{\tau}(\varepsilon))_{\varepsilon>0}$ with exponential distribution of parameter 1 on the same probability space $(\Omega, \mathscr{F}, \mathbb{P})$ as the driving Lévy noise $(L(t))_{t\geqslant 0}$ such that in probability*

$$\lim_{\varepsilon \to 0+} \inf_{y \in \hat{D}^{\pm}(\varepsilon^{\gamma})} |\lambda^{\pm}(\varepsilon)\tau_y^{\pm 0}(\varepsilon) - \bar{\tau}(\varepsilon)| = \lim_{\varepsilon \to 0+} \sup_{y \in \hat{D}^{\pm}(\varepsilon^{\gamma})} |\lambda^{\pm}(\varepsilon)\tau_y^{\pm 0}(\varepsilon) - \bar{\tau}(\varepsilon)| = 0.$$

6.2 Transition Times Between Balls Centered in the Stable States

In this section we shall investigate the asymptotic behavior in the small noise limit $\varepsilon \to 0$ of the times needed to switch between small neighborhoods of the metastable states. As in the Gaussian case it turns out that they differ only to a negligible extent from the transition or exit times investigated before.

We are next interested in times needed to transit from small neighborhoods of stable states $\phi^{\pm}$ to small neighborhoods of the opposite stable state $\phi^{\mp}$. A consequence of the Theorem 6.3 is that, starting in a small ball around $\phi^{\pm}$, the time needed to enter a small ball around the opposite stable state is of the order of the first exit time. Since the result in Proposition 4.5 contains a statement that holds only in probability, we cannot expect a convergence result for transition times as strong as the result of Theorem 5.8 for exit times. Instead, since the result in Proposition 4.5 is only in probability we cannot expect anything stronger (Fig. 6.1).

Let $\chi_x^{\pm}(\varepsilon)$ describe the time needed for X^{ε} starting in $t = 0$ at $X^{\varepsilon}(0) = x$ to enter the ball $B_{\varepsilon^{2\gamma}}(\phi^{\mp})$ contained in the *opposite* domain of attraction.

Definition 6.5. For $x \in \hat{D}^{\pm}(\varepsilon^{\gamma})$ we define the *first entrance time of a neighborhood of the opposite stable state* $\phi^{\pm}$

$$\chi_x^{\pm}(\varepsilon) := \inf\{t > 0 \mid X^{\varepsilon}(t; x) \in B_{\varepsilon^{2\gamma}}(\phi^{\mp})\}.$$

Remark 6.6. Note that by definition $X^{\varepsilon}(\tau^{\pm,0}(\varepsilon); x) \in \hat{D}^{\mp}(\varepsilon^{\gamma})$ for all $x \in \hat{D}^{\pm}(\varepsilon^{\gamma})$, which differs clearly from $\chi_x^{\pm}(\varepsilon)$ for systems with more than two stable solutions.

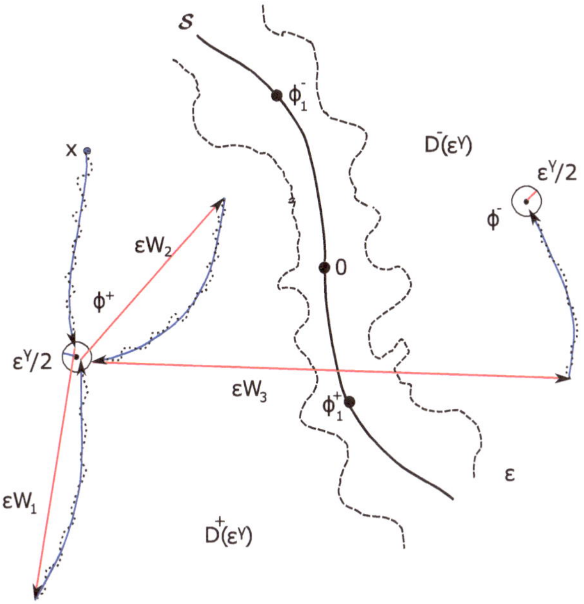

Fig. 6.1 Sketch of a transition event

The following proposition confirms that in our situation the asymptotic transitions between small neighborhoods of the stable states of the deterministic Chafee–Infante equation do not differ essentially from the asymptotic first exit times from reduced domains of attraction.

Theorem 6.7 (Asymptotic transitions between balls around the stable states). *Assume that (H.1) and (H.2) and Convention (C) are satisfied. Then there is a family of random variables* $(\bar{\tau}(\varepsilon))_{\varepsilon>0}$ *with exponential law of parameter 1 and* $h_0 > 0$ *such that for any* $0 < h \leqslant h_0$

$$\lim_{\varepsilon\to 0+} \mathbb{E}\left[\sup_{x\in\hat{D}^{\pm}(\varepsilon^{\gamma})} \mathbf{1}\{|\lambda^{\pm}(\varepsilon)\chi_x^{\pm}(\varepsilon) - \bar{\tau}(\varepsilon)| > h\}\right] = 0.$$

Proof. By Lemma 2.10 for any $x \in D^{\pm}$ there is $\varepsilon_0 > 0$ such that for $0 < \varepsilon \leqslant \varepsilon_0$ we obviously have $\hat{D}(\varepsilon^{\gamma})$ and

$$\tau_x^{\pm}(\varepsilon) \leqslant \tau_x^{\pm 0}(\varepsilon) \leqslant \chi_x^{\pm}(\varepsilon) \qquad \mathbb{P}-\text{a.s.} \tag{6.6}$$

since inequality (6.6) can actually be rewritten as

$$\begin{aligned}
&\inf\{t > 0 \mid X^\varepsilon(t;x) \in \big(\tilde{D}^\pm(\varepsilon^\gamma)\big)^c\} \\
&\quad \leqslant \inf\{t > 0 \mid X^\varepsilon(t;x) \in \Big(\tilde{D}^\pm(\varepsilon^\gamma) \cup \hat{D}^0(\varepsilon^\gamma)\Big)^c\} \\
&\quad \leqslant \inf\{t > 0 \mid X^\varepsilon(t;x) \in \Big(\tilde{D}^\pm(\varepsilon^\gamma) \cup \hat{D}^0(\varepsilon^\gamma) \cup (\hat{D}^\mp(\varepsilon^\gamma) \setminus B_{\varepsilon^{2\gamma}}(\phi^\mp))\Big)^c\}.
\end{aligned}$$

We now recall the definition (4.3) of the event E_x and define for $T \geqslant 0$ and $x \in H$ in a similar way

$$E_x^T := \mathscr{E}_{\tau^*}(\varepsilon^{2\gamma}) \cap \{ \sup_{s\in[\tau^*,T]} \|Y^\varepsilon(s;x) - u(s-\tau^*; Y^\varepsilon(\tau^*;x))\| \leqslant (1/2)\varepsilon^{2\gamma} \},$$

where $\tau^* = T \wedge s_{r^*}$. We further recall the path shift θ_t of $t > 0$. For convenience we introduce the notation

$$\begin{aligned}
E_x^T \circ \theta_t := \{ &\sup_{s\in[t,t+\tau^*]} \|\varepsilon\xi^*(t)\| \leqslant \varepsilon^{2\gamma}\} \\
&\cap \{ \sup_{s\in[t+\tau^*,t+T]} \|Y^\varepsilon(s;x) - u(s-\tau^*-t; Y^\varepsilon(\tau^*+t;x))\| \leqslant (1/2)\varepsilon^{2\gamma} \},
\end{aligned}$$

for $\tau^* = s_{r^*} \wedge T$ and $T \geqslant 0$. Clearly $E_x^T = E_x^T \circ \theta_0$. In an analogous way we define $\sigma_{r^*}(\cdot,\varepsilon) \circ \theta_T$.

Claim:

For $\kappa > 0$ we have

$$\lim_{\varepsilon\to 0+} \mathbb{E}\left[\sup_{x\in\hat{D}^\pm(\varepsilon^\gamma)} \mathbf{1}\{\chi_x^\pm(\varepsilon) < \tau_x^{\pm 0}(\varepsilon) + \sigma_{r^*}(\cdot,\varepsilon) \circ \theta_{\tau_x^{\pm 0}(\varepsilon)} + T_{rec} + \kappa\gamma|\ln\varepsilon|\} \right] = 1. \tag{6.7}$$

To prove the claim, fix $\varepsilon, \gamma, \kappa > 0$. We rely on the strong Markov property of our process X^ε and the stopping times $\tau_x^{\pm 0}(\varepsilon)$ and $\sigma_{r^*}(\cdot,\varepsilon) \circ \theta_{\tau_x^{\pm 0}(\varepsilon)}$, see (4.2). To lighten notation we write τ_ε instead of $\tau_x^{\pm 0}(\varepsilon) + \sigma_{r^*}(\cdot,\varepsilon) \circ \theta_{\tau_x^{\pm 0}(\varepsilon)}$ and T^ε instead of $T_{rec} + \kappa\gamma|\ln\varepsilon|$.

We have

$$\mathbb{E}\left[\sup_{x\in\hat{D}^{\pm}(\varepsilon^{\gamma})}\mathbf{1}\{\chi_x^{\pm}(\varepsilon)\geqslant\tau_\varepsilon+T^\varepsilon\}\right]$$
$$\leqslant\mathbb{E}\left[\sup_{x\in\hat{D}^{\pm}(\varepsilon^{\gamma})}\mathbf{1}\{\|X^\varepsilon(\tau_\varepsilon+T^\varepsilon;x)-\phi^{\mp}\|\geqslant\varepsilon^{2\gamma}\}\cap E_x^{T^\varepsilon}\circ\theta_{\tau_\varepsilon}\right]$$
$$+\mathbb{E}\left[\sup_{x\in\hat{D}^{\pm}(\varepsilon^{\gamma})}\mathbf{1}((E_x^{T^\varepsilon})^c)\right].$$

By Proposition 4.5 the latter term tends to zero. Moreover, by Propositions 4.7, 2.12, the Markov property and Remark 6.6

$$\mathbb{E}\left[\sup_{x\in\hat{D}^{\pm}(\varepsilon^{\gamma})}\mathbf{1}\{\|X^\varepsilon(\tau_\varepsilon+T^\varepsilon;x)-\phi^{\mp}\|\geqslant\varepsilon^{2\gamma}\}\mathbf{1}(E_x^{T^\varepsilon}\circ\theta_{\tau_\varepsilon})\right]$$
$$\leqslant\mathbb{E}\left[\sup_{x\in\hat{D}^{\pm}(\varepsilon^{\gamma})}\mathbf{1}\{\|X^\varepsilon(T^\varepsilon;X^\varepsilon(\tau_\varepsilon;x))-u(T^\varepsilon;X^\varepsilon(\tau_\varepsilon;x))\|\geqslant(1/2)\varepsilon^{2\gamma}\}\right]$$
$$+\;\mathbb{E}\left[\sup_{x\in\hat{D}^{\pm}(\varepsilon^{\gamma})}\mathbf{1}\{\|u(T^\varepsilon;X^\varepsilon(\tau_\varepsilon;x))-\phi^{\pm}\|\geqslant(1/2)\varepsilon^{2\gamma}\}\right]$$
$$\leqslant\mathbb{E}\left[\mathbf{1}\{\sup_{x\in\hat{D}^{\pm}(\varepsilon^{\gamma})}\sup_{t\in[0,T^\varepsilon]}\|X^\varepsilon(t;X^\varepsilon(\tau_\varepsilon;x))-u(t;X^\varepsilon(\tau_\varepsilon;x))\|\geqslant(1/2)\varepsilon^{2\gamma}\}\right]$$
$$\leqslant\mathbb{E}\left[\sup_{y\in\hat{D}^{\mp}(\varepsilon^{\gamma})\cap B_{r^*}(0)}\mathbf{1}\{\sup_{t\in[0,T^\varepsilon]}\|X^\varepsilon(t;y)-u(t;y)\|\geqslant(1/2)\varepsilon^{2\gamma}\}\right].$$

By Proposition 4.5 the latter term tends to zero. Further by Proposition 4.7, we find $\Gamma=\Gamma(\kappa)>0$ and $\varepsilon_0>0$ such that for $0<\varepsilon\leqslant\varepsilon_0$ and

$$\mathbb{E}\left[\sup_{y\in\hat{D}^{\mp}(\varepsilon^{\gamma})\cap B_{r^*}(0)}\mathbf{1}\{\sup_{t\in[0,T^\varepsilon]}\|X^\varepsilon(t;y)-u(t;y)\|\geqslant(1/2)\varepsilon^{2\gamma}\}\right]$$
$$\leqslant\mathbb{P}((\mathscr{E}_{T^\varepsilon}(\varepsilon^{(\Gamma+2)\gamma})\cap\{T_1\geqslant T^\varepsilon\})^c)\tag{6.8}$$
$$\leqslant\mathbb{P}(\mathscr{E}^c_{T^\varepsilon}(\varepsilon^{(\Gamma+2)\gamma}))+1-e^{-\beta_\varepsilon T^\varepsilon}.$$

Moreover, by Lemma 4.11 for ε_0 small enough and a constant $C>0$

$$\mathbb{P}(\mathscr{E}^c_{T^\varepsilon}(\varepsilon^{(\Gamma+2)\gamma}))\leqslant CT^\varepsilon\varepsilon^{2-2(\Gamma+2)\gamma-(2-(1-\Theta)\alpha)\rho},\tag{6.9}$$

such that the right hand side of the desired estimate converges to 0 as $\varepsilon \to 0+$. This establishes

$$\lim_{\varepsilon\to 0+} \mathbb{E}\left[\sup_{x\in\hat{D}^{\pm}(\varepsilon^{\gamma})} \mathbf{1}\{\chi_x^{\pm}(\varepsilon) \geqslant \tau_x^{\pm 0}(\varepsilon) + \sigma_{r^*}(\cdot,\varepsilon)\circ\theta_{\tau_x^{\pm 0}(\varepsilon)} + T^{\varepsilon}\}\right] = 0, \quad (6.10)$$

and therefore in combination with the inequalities (6.6) and (6.10)

$$\lim_{\varepsilon\to 0+} \mathbb{E}\Bigg[\inf_{x\in\hat{D}^{\pm}(\varepsilon^{\gamma})} \mathbf{1}\{\lambda^{\pm}(\varepsilon)\tau_x^{\pm 0}(\varepsilon) \leqslant \lambda^{\pm}(\varepsilon)\chi_x^{\pm}(\varepsilon)$$

$$\leqslant \lambda^{\pm}(\varepsilon)\left(\tau_x^{\pm 0}(\varepsilon) + \sigma_{r^*}(\cdot,\varepsilon)\circ\theta_{\tau_x^{\pm 0}(\varepsilon)} + T^{\varepsilon}\right)\}\Bigg] = 1. \quad (6.11)$$

By the strong Markov property of X^{ε} and ξ^{ε} and Proposition 4.7 we have

$$\sigma_{r^*}(\cdot,\varepsilon)\circ\theta_{\tau_x^{\pm 0}(\varepsilon)} \leqslant s_{r^*}$$

$$\mathbb{P}(\ \cdot\ |\ \mathscr{E}_{s_{r^*}}(\varepsilon^{(\Gamma+2)\gamma})\circ\theta_{\tau_x^{\pm 0}(\varepsilon)} \cap \{T_1\circ\theta_{\tau_x^{\pm 0}(\varepsilon)} \geqslant s_{r^*}+T^{\varepsilon}\})\text{-a. s.},$$

where clearly $\mathbb{P}(T_1\circ\theta_{\tau_x^{\pm 0}(\varepsilon)} \geqslant s_{r^*}+T^{\varepsilon}) = \mathbb{P}(T_1 \geqslant s_{r^*}+T^{\varepsilon}) = e^{-\beta_{\varepsilon}(s_{r^*}+T^{\varepsilon})} \to 1$ and

$$\lim_{\varepsilon\to 0+} \mathbb{P}(\mathscr{E}_{s_{r^*}}(\varepsilon^{(\Gamma+2)\gamma})\circ\theta_{\tau_x^{\pm 0}(\varepsilon)}) = \lim_{\varepsilon\to 0+} \mathbb{P}(\mathscr{E}_{s_{r^*}}(\varepsilon^{(\Gamma+2)\gamma})) = 1.$$

With this in mind we recall that by Corollary 6.4 there is a family of random variables $(\bar{\tau}(\varepsilon))_{\varepsilon>0}$ with exponential law of parameter 1 such that

$$\lim_{\varepsilon\to 0+} \sup_{x\in\hat{D}^{\pm}(\varepsilon)} |\lambda^{\pm}(\varepsilon)\,(\tau_{\varepsilon}+T^{\varepsilon}) - \bar{\tau}(\varepsilon)| = \lim_{\varepsilon\to 0+} \sup_{x\in\hat{D}^{\pm}(\varepsilon)} |\lambda^{\pm}(\varepsilon)\tau_x^{\pm 0}(\varepsilon) - \bar{\tau}(\varepsilon)| = 0$$

in probability. This implies for given $h > 0$

$$\lim_{\varepsilon\to 0+} \mathbb{E}\left[\sup_{x\in\hat{D}^{\pm}(\varepsilon)} \mathbf{1}\{|\lambda^{\pm}(\varepsilon)\chi_x^{\pm 0}(\varepsilon) - \bar{\tau}(\varepsilon)| > h\}\right]$$

$$\leqslant \lim_{\varepsilon\to 0+} \mathbb{P}\left(\sup_{x\in\hat{D}^{\pm}(\varepsilon)} |\lambda^{\pm}(\varepsilon)\chi_x^{\pm 0}(\varepsilon) - \bar{\tau}(\varepsilon)| > h\right) = 0.$$

This finishes the proof. □

Chapter 7
Localization and Metastability

In this chapter, equipped with our previously obtained knowledge of exit and transition times in the limit of small noise amplitude $\varepsilon \to 0$, we shall investigate the global asymptotic behavior of our jump diffusion process in the time scale in which transitions occur, i.e. in the scale given by $\lambda^0(\varepsilon) = \nu(\frac{1}{\varepsilon} B_\delta^c(0)), \varepsilon, \delta > 0$. It turns out that in this time scale, the switching of the diffusion between neighborhoods of the stable solutions $\phi^\pm$ can be well described by a Markov chain jumping back and forth between two states with a characteristic Q-matrix determined by the quantities $\frac{\mu((D_0^\pm)^c)}{\mu(B_\delta^c(0))}$ as jumping rates. To show this, we need to prove that X^ε is localized around the stable fixed points also on the *critical* time scale $T/\lambda^0(\varepsilon)$. This boils down to the control of the exit behavior from the complement of the reduced domains of attraction $\hat{D}^0(\varepsilon^\gamma)$. Roughly, rates for large jumps between positions inside this set have to converge to 0 with $\varepsilon \to 0$. This condition is made precise in Hypothesis (H.3), that plays an important role in the subsequent study of metastability.

7.1 Hypothesis (H.3) Prevents Trapping Close to the Separatrix

Recall that by Proposition 4.1 there is r^* and s_{r^*} such that

$$\|u(t;x)\| \leqslant r^* \qquad t \geqslant s_{r^*}, x \in H.$$

Clearly this remains true for all $r \geqslant r^*$ and respective $s_r \leqslant s_{r^*}$.

We introduce after (H.1) and (H.2) a third non-degeneracy for the Lévy measure with respect to the separatrix $\mathscr{S}$.

(H.3) Restriction on large jumps close to the separatrix: *Let γ be given according to Conventions (C). Let there be $\gamma/2 < \hat{\gamma} \leqslant \gamma$ and $r \geqslant r^*$ given by Proposition 4.1 such that*

A. Debussche et al., *The Dynamics of Nonlinear Reaction-Diffusion Equations with Small Lévy Noise*, Lecture Notes in Mathematics 2085, DOI 10.1007/978-3-319-00828-8_7,

$$\lim_{\varepsilon\to 0+} \sup_{x\in B_r(0)\cap \hat D^0(\varepsilon^\gamma)} \frac{\nu\left(\frac{1}{\varepsilon^{1-\hat\gamma}} B_1^c(0)\cap \frac{1}{\varepsilon}\left(\hat D^0(\varepsilon^\gamma)-x\right)\right)}{\nu\left(\frac{1}{\varepsilon^{1-\hat\gamma}} B_1^c(0)\right)} = 0.$$

This condition stipulates that the probability for large jumps from positions inside $\hat D^0(\varepsilon^\gamma)$ bounded by r to $\hat D^0(\varepsilon^\gamma)$ itself tends to zero with ε for some parameter $r > r^*$.

In this section we shall justify Hypothesis (H.3) with an explanation of the precise choice of $\hat\gamma$ and prove crucial implications.

Definition 7.1. For $\varepsilon > 0$ and $X^\varepsilon(\cdot; x)$ the càdlàg mild solution of (1.2), $\gamma \in (0,1)$, and $x \in \hat D^0(\varepsilon^\gamma)$ let

$$\tau_x^0(\varepsilon) := \inf\{t > 0 \mid X^\varepsilon(t;x) \notin \hat D^0(\varepsilon^\gamma)\}$$

be the *first exit time from the neighborhood of the separatrix* $\hat D^0(\varepsilon^\gamma)$.

Definition 7.2. We assume Hypotheses (H.1), (H.2) and (H.3) and Convention (C) to be satisfied. For the constant $\gamma/2 < \hat\gamma \leqslant \gamma$ from (H.3), $\varepsilon > 0$, set $\hat\beta_\varepsilon := \nu\left(\frac{1}{\varepsilon^{1-\hat\gamma}} B_1^c(0)\right)$, and recursively for $i \in \mathbb{N}$

$$\hat T_0 := 0, \qquad \hat T_{i+1} := \inf\{t > T_i \mid \varepsilon\|\Delta_t L\| > \varepsilon^{\hat\gamma}\}, \qquad \text{and} \qquad \hat W_i := \Delta_{\hat T_i} L. \tag{7.1}$$

By definition, $\hat T_1$ then has an exponential law with parameter $\hat\beta_\varepsilon$. In analogy to Sect. 3.1 we shall split $L = \hat\eta^\varepsilon + \hat\xi^\varepsilon$, where

$$\hat\eta^\varepsilon(t) = \sum_{T_i \leqslant t} \hat W_i, \qquad t \geqslant 0,$$

is a compound Poisson process with jump probability measure $\frac{1}{\hat\beta_\varepsilon}\nu(\cdot \cap \frac{1}{\varepsilon^{1-\hat\gamma}} B_1^c(0))$. For further use we denote by

$$\hat\xi^*(t) := \int_0^t S(t-s)\, \mathrm{d}\hat\xi^\varepsilon(s), \qquad t \geqslant 0,$$

and $\hat Y^\varepsilon$ the solution of (3.11) driven by $\varepsilon\hat\xi^\varepsilon$ instead of $\varepsilon\xi^\varepsilon$. We recall the notation

$$\lambda^0(\varepsilon) = \nu\left(\frac{1}{\varepsilon} B_1^c(0)\right) = \varepsilon^\alpha \ell(1/\varepsilon)\mu(B_1^c(0))$$

with the slowly varying function ℓ.

Remark 7.3.

1. Let us argue why we introduce the additional parameter $\hat{\gamma} \in (\gamma/2, \gamma]$. The upper and the lower bounds are derived from two properties important in the sequel.

1.1 In Lemma 7.4 we shall compare $\tau_x^0(\varepsilon)$ with a deterministic time scale $\frac{1}{\varepsilon^g}$, where $g > 0$. In the proof this reduces to the comparison of $\hat{T}_1$ with $\frac{1}{\varepsilon^g}$. We aim at showing that $\lim_{\varepsilon \to 0+} \mathbb{P}(\hat{T}_1 > \frac{1}{\varepsilon^g}) = 0$. For this purpose we may calculate

$$\mathbb{P}(\hat{T}_1 > \frac{1}{\varepsilon^g}) = \int_0^{\frac{1}{\varepsilon^g}} \hat{\beta}_\varepsilon \exp\left(-\hat{\beta}_\varepsilon s\right) \, ds = 1 - \exp\left(-\frac{\hat{\beta}_\varepsilon}{\varepsilon^g}\right)$$

$$= 1 - \exp\left(-\frac{\varepsilon^{\alpha(1-\hat{\gamma})} \ell\left(\frac{1}{\varepsilon}\right) \mu(B_1^c(0))}{\varepsilon^g}\right).$$

In order that the last term tends to zero it is sufficient that $g > \alpha(1 - \hat{\gamma})$. This follows for $g = \alpha(1 - \gamma/2)$, if $\hat{\gamma} > \gamma/2$.

1.2 In Lemma 7.5 we shall compare $\tau_x^0(\varepsilon)$ with $\hat{T}_1$ with the aim to prove $\lim_{\varepsilon \to 0+} \mathbb{P}(\tau_x^0(\varepsilon) > \hat{T}_1) = 0$. By definition $\operatorname{dist}(\hat{D}^\pm(\varepsilon^\gamma), \mathscr{S}) > \varepsilon^\gamma$. Therefore the choice of $\hat{\gamma}$ must not inhibit that $\Delta_{\hat{T}_i} X^\varepsilon = \varepsilon \Delta_{\hat{T}_i} L = \varepsilon \hat{W}_i$ for $i \in \mathbb{N}$ triggers the exit of X^ε from $\hat{D}^0(\varepsilon^\gamma)$ with a reasonable chance. Hence for any reasonable choice of $\hat{\gamma}$ there must be a constant $C > 0$ such that at least for small $\varepsilon > 0$

$$\mathbb{P}(\|\varepsilon \hat{W}_i\| > \varepsilon^\gamma + \varepsilon^{2\gamma}) > C.$$

Assume $\hat{\gamma} > \gamma$ and $y \in \mathscr{S}$. Then

$$\mathbb{P}(y + \varepsilon \hat{W}_i \in \hat{D}^+(\varepsilon^\gamma) \cup \hat{D}^-(\varepsilon^\gamma)) = \frac{\nu\left(\frac{1}{\varepsilon^{1-\hat{\gamma}}}(B_1^c(0)) \cap \frac{1}{\varepsilon}(\hat{D}^0(\varepsilon^\gamma))^c\right)}{\nu\left(\frac{1}{\varepsilon^{1-\hat{\gamma}}} B_1^c(0)\right)}$$

$$\leqslant \frac{\nu\left(\frac{1}{\varepsilon^{1-\hat{\gamma}}} B_1^c(0) \cap \frac{1}{\varepsilon^{1-\gamma}} B_1^c(0)\right)}{\nu\left(\frac{1}{\varepsilon^{1-\hat{\gamma}}} B_1^c(0)\right)}$$

$$= \frac{\ell(\frac{1}{\varepsilon^{1-\gamma}})}{\ell(\frac{1}{\varepsilon^{1-\hat{\gamma}}})} \varepsilon^{\hat{\gamma}-\gamma} \to 0,$$

as $\varepsilon \to 0+$. In other words $\varepsilon \hat{W}_i$ is of an asymptotically too small scale to trigger an exit from $\hat{D}^0(\varepsilon^\gamma)$. Hence $\hat{\gamma} \leqslant \gamma$ is a necessary condition. Apart from the parameter $r \geqslant r^*$, Hypothesis (H.3) on ν appears now quite natural, since it can be paraphrased as stating that there is $\gamma/2 < \hat{\gamma} \leqslant \gamma$ such that

$$\sup_{y \in B_r(0) \cap \hat{D}^0(\varepsilon^\gamma)} \mathbb{P}\left(y + \varepsilon \hat{W}_1 \in \hat{D}^+(\varepsilon^\gamma) \cup \hat{D}^-(\varepsilon^\gamma)\right)$$

$$= 1 - \sup_{y \in B_r(0) \cap \hat{D}^0(\varepsilon^\gamma)} \mathbb{P}\left(y + \varepsilon \hat{W}_1 \in \hat{D}^0(\varepsilon^\gamma)\right) \to 1, \text{ as } \varepsilon \to 0+ .$$

This means that by large jumps $\varepsilon \hat{W}_i$ of size at least $\varepsilon^{\hat{\gamma}}$ the process should have a non-negligible chance to leave the set $\hat{D}^0(\varepsilon^\gamma)$.

2. Could Hypothesis (H.3) be slightly strengthened, to improve the metastability results of Theorems 7.10 and 7.11? For instance, could the supremum over $x \in H$ appearing in Lemma 7.8 or the one over $x \in D^\pm$ in Theorem 7.10 be taken under the expectation sign, by just requiring a slightly stronger hypothesis? Let us argue that the hypothesis

$$\lim_{\varepsilon \to 0+} \mathbb{E}\left[\sup_{y \in B_r(0) \cap \hat{D}^0(\varepsilon^\gamma)} \mathbf{1}\{y + \varepsilon \hat{W}_1 \in \hat{D}^+(\varepsilon^\gamma) \cup \hat{D}^-(\varepsilon^\gamma)\}\right]$$

$$= \lim_{\varepsilon \to 0+} \frac{\nu\left(\frac{1}{\varepsilon^{1-\hat{\gamma}}} B_1^c(0) \cap \frac{1}{\varepsilon}\left(\hat{D}^0(\varepsilon^\gamma) - \left(B_r(0) \cap \tilde{D}^0(\varepsilon^\gamma)\right)\right)\right)}{\hat{\beta}_\varepsilon} = 0,$$

slightly stronger than Hypothesis (H.3), is in general too strong for purely geometric reasons. Recalling Lemma 2.10 the second claim of which states

$$\bigcap_{\varepsilon > 0} \hat{D}^0(\varepsilon^\gamma) = \mathscr{S},$$

we can say that apart from very special cases (for example in case $\mathscr{S}$ is contained in a subspace of codimension 1), the separatrix $\mathscr{S}$ will not be contained in

$$\bigcap_{\varepsilon > 0} \hat{D}^0(\varepsilon^\gamma) - \left(B_r(0) \cap \hat{D}^0(\varepsilon^\gamma)\right) = \mathscr{S} - (B_r(0) \cap \mathscr{S}) .$$

In this case there generically exists a small ball $B_h(0) \subset \mathscr{S} - (B_r(0) \cap \mathscr{S})$. But then

$$\frac{\mu\left(B_1^c(0)\cap\frac{1}{\varepsilon^{\hat\gamma}}\left(\tilde D^0(\varepsilon^\gamma)-\left(B_r(0)\cap\hat D^0(\varepsilon^\gamma)\right)\right)\right)}{\mu\left(B_1^c(0)\right)}$$
$$\geqslant\frac{\mu\left(B_1^c(0)\cap\frac{1}{\varepsilon^{\hat\gamma}}\left(\mathscr{S}-\left(B_r(0)\cap\mathscr{S}\right)\right)\right)}{\mu\left(B_1^c(0)\right)}$$
$$\geqslant\frac{\mu\left(B_1^c(0)\cap\frac{1}{\varepsilon^{\tilde\gamma}}\left(B_h(0)\right)\right)}{\mu\left(B_1^c(0)\right)}\overset{\varepsilon\to0+}{\longrightarrow}1, \tag{7.2}$$

and

$$\frac{\nu\left(\frac{1}{\varepsilon^{1-\hat\gamma}}B_1^c(0)\cap\frac{1}{\varepsilon}\left(\hat D^0(\varepsilon^\gamma)-\left(B_r(0)\cap\hat D^0(\varepsilon^\gamma)\right)\right)\right)}{\hat\beta_\varepsilon}$$
$$\left[\frac{\mu\left(B_1^c(0)\cap\frac{1}{\varepsilon^{\hat\gamma}}\left(\tilde D^0(\varepsilon^\gamma)-\left(B_r(0)\cap\hat D^0(\varepsilon^\gamma)\right)\right)\right)}{\mu\left(B_1^c(0)\right)}\right]^{-1}\to1,\quad \text{as } \varepsilon\to0+,$$

and therefore

$$\frac{\nu\left(\frac{1}{\varepsilon^{1-\hat\gamma}}B_1^c(0)\cap\frac{1}{\varepsilon}\left(\hat D^0(\varepsilon^\gamma)-\left(B_r(0)\cap\hat D^0(\varepsilon^\gamma)\right)\right)\right)}{\hat\beta_\varepsilon}\to1,\quad \text{as } \varepsilon\to0+\,.$$

It is therefore difficult to find a stronger hypothesis enhancing the quality of convergence in our metastability results.

The main consequence of Hypothesis (H.3) that will be vastly exploited in the sequel is contained in the following lemma.

Lemma 7.4. *Suppose the Hypotheses (H.1), (H.2) and (H.3) and Convention (C) are satisfied. Then for any $T>0$*

$$\lim_{\varepsilon\to0+}\sup_{x\in\hat D^0(\varepsilon^\gamma)}\mathbb{P}\left(\tau_x^0(\varepsilon)>\frac{T}{\varepsilon^{\alpha(1-\gamma/2)}}\right)=0$$

holds true.

Proof. In fact, for $\varepsilon>0$ and the first jump $\hat T_1$ of the compound Poisson process $\hat\eta^\varepsilon$ we have

$$\sup_{x\in\hat{D}^0(\varepsilon^\gamma)} \mathbb{P}\left(\tau_x^0(\varepsilon) > \frac{T}{\varepsilon^{\alpha(1-\gamma/2)}}\right)$$

$$\leqslant \sup_{x\in\hat{D}^0(\varepsilon^\gamma)} \mathbb{P}\left(\{\tau_x^0(\varepsilon) > \frac{T}{\varepsilon^{\alpha(1-\gamma/2)}}\} \cap \{\tau_x^0(\varepsilon) \leqslant \hat{T}_1\}\right)$$

$$+ \sup_{x\in\hat{D}^0(\varepsilon^\gamma)} \mathbb{P}\left(\tau_x^0(\varepsilon) > \frac{T}{\varepsilon^{\alpha(1-\gamma/2)}} \mid \tau_x^0(\varepsilon) > \hat{T}_1\right) \mathbb{P}\left(\tau_x^0(\varepsilon) > \hat{T}_1\right)$$

$$\leqslant \sup_{x\in\hat{D}^0(\varepsilon^\gamma)} \mathbb{P}\left(\hat{T}_1 > \frac{T}{\varepsilon^{\alpha(1-\gamma/2)}}\right) + \sup_{x\in\hat{D}^0(\varepsilon^\gamma)} \mathbb{P}\left(\tau_x^0(\varepsilon) > \hat{T}_1\right)$$

$$\leqslant \exp\left(-\frac{T\,\hat{\beta}_\varepsilon}{\varepsilon^{\alpha(1-\gamma/2)}}\right) + \sup_{x\in\hat{D}^0(\varepsilon^\gamma)} \mathbb{P}\left(\tau_x^0(\varepsilon) > \hat{T}_1\right).$$

Since $\hat{\gamma} > \gamma/2$, we obtain $\lim_{\varepsilon\to 0+} \hat{\beta}_\varepsilon/\varepsilon^{\alpha(1-\gamma/2)} = \infty$, such that the first term in the last line of the preceding estimate tends to zero. The second term

$$p_1(\varepsilon) := \sup_{y\in\hat{D}^0(\varepsilon^\gamma)} \mathbb{P}\left(\tau_y^0(\varepsilon) > \hat{T}_1\right), \varepsilon > 0,$$

will be estimated in the subsequent Lemma 7.5. □

Lemma 7.5. *Under the assumptions of Lemma 7.4 the relationship*

$$\lim_{\varepsilon\to 0+} \sup_{x\in\hat{D}^0(\varepsilon^\gamma)} \mathbb{P}\left(\tau_x^0(\varepsilon) > \hat{T}_1\right) = 0$$

is satisfied.

Proof. For $r > 0$, $\varepsilon > 0$ and $x \in \hat{D}^0(\varepsilon^\gamma)$ we may write

$$\mathbb{P}\left(\tau_x^0(\varepsilon) > \hat{T}_1\right)$$

$$= \mathbb{P}\left(\tau_x^0(\varepsilon) > \hat{T}_1 \text{ and } X^\varepsilon(\hat{T}_1-;x) \in B_r(0) \cap \hat{D}^0(\varepsilon^\gamma)\right)$$

$$+ \mathbb{P}\left(\tau_x^0(\varepsilon) > \hat{T}_1 \text{ and } X^\varepsilon(\hat{T}_1-;x) \notin B_r(0) \cap \hat{D}^0(\varepsilon^\gamma)\right)$$

$$\leqslant \mathbb{P}\left(\hat{Y}^\varepsilon(\hat{T}_1;x) \in B_r(0) \cap \hat{D}^0(\varepsilon^\gamma) \text{ and } \hat{Y}^\varepsilon(\hat{T}_1,x) + \varepsilon\hat{W}_1 \in \hat{D}^0(\varepsilon^\gamma)\right)$$

$$+ \mathbb{P}\left(\hat{Y}^\varepsilon(\hat{T}_1;x) \notin B_r(0)\right)$$

$$\leqslant \sup_{y \in B_r(0) \cap \hat{D}^0(\varepsilon^\gamma)} \mathbb{P}\left(y + \varepsilon \hat{W}_1 \in \hat{D}^0(\varepsilon^\gamma)\right) + \mathbb{P}\left(\sup_{x \in H} \|\hat{Y}^\varepsilon(\hat{T}_1; x)\| > r\right).$$

By Hypothesis (H.3) we may treat the first term by noting

$$\sup_{y \in B_r(0) \cap \hat{D}^0(\varepsilon^\gamma)} \mathbb{P}\left(y + \varepsilon \hat{W}_1 \in \hat{D}^0(\varepsilon^\gamma)\right)$$

$$= \sup_{y \in B_r(0) \cap \hat{D}^0(\varepsilon^\gamma)} \frac{\nu\left(\frac{1}{\varepsilon^{1-\hat{\gamma}}} B_1^c(0) \cap \frac{1}{\varepsilon}\left(\hat{D}^0(\varepsilon^\gamma) - y\right)\right)}{\hat{\beta}_\varepsilon} \overset{\varepsilon \to 0+}{\longrightarrow} 0. \tag{7.3}$$

In order to show the convergence of the second term we use the representation of $\hat{Y}^\varepsilon(t; x) = v_{\varepsilon\hat{\xi}^*}(t; x) + \varepsilon\hat{\xi}^*(t)$. Proposition 4.1 states the existence of $r^* > 0$ such that for all $t \geqslant s_{r^*}$

$$\sup_{x \in H} \|\hat{Y}^\varepsilon(t; x)\| \leqslant r^*$$

on the event $\{\sup_{t \geq 0} \|\varepsilon\hat{\xi}^*(t)\| \leqslant 1\}$. This implies for any $r \geqslant r^*$ a constant $s_r \leqslant s_{r^*}$ with the same relation. Hence we may calculate directly

$$\mathbb{P}\left(\sup_{x \in H} \|\hat{Y}^\varepsilon(\hat{T}_1; x)\| > r\right) = \int_0^{s_r} + \int_{s_r}^\infty \mathbb{P}\left(\sup_{x \in H} \|\hat{Y}^\varepsilon(s; x)\| > r\right) \hat{\beta}_\varepsilon e^{-\hat{\beta}_\varepsilon s} ds$$

$$\leqslant s_r \hat{\beta}_\varepsilon + \int_{s_r}^\infty \mathbb{P}\left(\sup_{x \in H} \|\hat{Y}^\varepsilon(s; x)\| > r \mid \sup_{\sigma \in [s_r, s]} \|\varepsilon\hat{\xi}^*(\sigma)\| \leqslant 1\right)$$

$$\mathbb{P}(\sup_{\sigma \in [s_r, s]} \|\varepsilon\hat{\xi}^*(\sigma)\| \leqslant 1) \hat{\beta}_\varepsilon e^{-\hat{\beta}_\varepsilon s} ds$$

$$+ \int_{s_r}^\infty \mathbb{P}(\sup_{\sigma \in [s_r, s]} \|\varepsilon\hat{\xi}^*(\sigma)\| \geqslant 1) \hat{\beta}_\varepsilon e^{-\hat{\beta}_\varepsilon s} ds.$$

The second term vanishes and the two others converges to 0 as $\varepsilon \to 0+$. In particular for the third one, one may use the arguments used in Sects. 4.3.2 and 4.4. □

7.2 Localization on Subcritical and Critical Time Scales

In Chaps. 5 and 6 we have seen that exits and transitions between relevant areas in the domains of attraction of the metastable solutions are of the order of $\varepsilon^{-\alpha}$. We shall now consider our system on time scales smaller than this threshold. We shall thereby confirm the reasonable conjecture that on these time scales the solution trajectories converge in probability to the process spending all the time at the local minimum $\phi^{\pm}$ associated to the domain of attraction of the starting value $x \in D^{\pm}$. Our result is even stronger. We prove that after an initial relaxation time of the order of magnitude $s_{r^*} + T_{rec} + \kappa\gamma|\ln\varepsilon|$, the solution trajectories of the stochastic Chafee–Infante equation converge to the deterministic stationary solutions in $\phi^{\pm}$ uniformly in probability.

If $0 < \delta < \alpha$ and we consider the entire process $(X^{\varepsilon}(t/\varepsilon^{\delta}))_{t\in[0,T]}$ for fixed $T > 0$ it should converge for $\varepsilon \to 0+$ to the process taking the constant value given by the stable state in the domain of attraction where it started. This can be justified, since the relaxation time of order $s_{r^*} + T_{rec} + \kappa\gamma|\ln\varepsilon|$ of the small jump solution Y^{ε} of (3.11) to the stable state is clearly dominated by $\frac{1}{\varepsilon^{\delta}}$, but the first exit time $\tau_x^{\pm}(\varepsilon)$ of expected order $\frac{1}{\lambda^{\pm}(\varepsilon)} \approx \frac{1}{\lambda^{0}(\varepsilon)} \approx \frac{1}{\varepsilon^{\alpha}}$ is not yet reached. However, this is only true if we avoid initial values close to the separatrix $\mathscr{S}$. This is the infinite-dimensional analogon to a result in [IP08].

Theorem 7.6 (Uniform convergence in probability on time scales). *Assume that Hypotheses (H.1) and (H.2) and Convention (C) are satisfied and $T_{rec}, \kappa, s_{r^*} > 0$ are given by Proposition 2.12 and 4.1. Fix $0 < \delta < \alpha$. Then there is $h_0 > 0$ such that for $0 < h \leqslant h_0$ and for any $T > 0$*

$$\lim_{\varepsilon\to 0+} \mathbb{E}\left[\sup_{x\in\hat{D}^{\pm}(\varepsilon^{\gamma})}\ \sup_{t\in[s_{r^*}+T_{rec}+\kappa\gamma|\ln\varepsilon|, T/\varepsilon^{\delta}]} \mathbf{1}\{\|X^{\varepsilon}(t;x) - \phi^{\pm}\| > h\}\right] = 0. \tag{7.4}$$

Proof. Lemma 2.13 states the existence of constants $\delta_s > 0$ and $C_2 > 0$ such that for all $0 < \sigma < \delta_s$, $y \in B_{\sigma}(\phi^{\pm})$ and $t \geqslant 0$

$$u(t;y) \in B_{C_2\sigma}(\phi^{\pm}).$$

With $\Gamma = \Gamma(\kappa) > 0$ according to Proposition 4.7 and $s_{r^*} > 0$ given by Proposition 4.1 we define the event

$$\tilde{\mathscr{E}}(\varepsilon) := \{\sup_{t\in[0,s_{r^*}+T_{rec}+\kappa\gamma|\ln\varepsilon|]} \|\varepsilon\xi^*(t)\| < \varepsilon^{(\Gamma+2)\gamma}\} \cap \{T_1 > s_{r^*} + T_{rec} + \kappa\gamma|\ln\varepsilon|\}.$$

We can now estimate by the supremum over all possible states the process may take after $s_{r^*} + T_{rec} + \kappa\gamma|\ln\varepsilon|$ time units. We use the Markov property of X^{ε} and Y^{ε} in two steps. First we exploit Proposition 4.1, which yields that for times beyond s_{r^*} the process Y^{ε} stays within the ball $B_{r^*}(0)$. Secondly, Proposition 4.7 for initial values in of Y^{ε} in $B_{r^*}(0)$. Hence we obtain for $0 < h \leqslant h_0 := C_2\delta_s$ and $0 < \varepsilon \leqslant \varepsilon_0$ with at least $\varepsilon_0^{2\gamma} < \delta_s$ but small enough such that all involved times are nonnegative and the above mentioned results are true

$$\mathbb{E}\left[\sup_{x\in\hat{D}^{\pm}(\varepsilon^{\gamma})}\ \sup_{t\in[s_{r^*}+T_{rec}+\kappa\gamma|\ln\varepsilon|, T/\varepsilon^{\delta}]} \mathbf{1}\{\|X^{\varepsilon}(t;x) - \phi^{\pm}\| \geqslant h\}\right] \tag{7.5}$$

$$= \mathbb{E}\Big[\sup_{x\in\hat{D}^\pm(\varepsilon^\gamma)} \mathbf{1}\{ \sup_{t\in[T_{rec}+\kappa\gamma|\ln\varepsilon|, T/\varepsilon^\delta - s_r*]} \|X^\varepsilon(t;x)\circ\theta_{s_r*} - \phi^\pm\| \geqslant h\}$$

$$\cdot \left(\mathbf{1}(\tilde{\mathscr{E}}(\varepsilon)) + \mathbf{1}(\tilde{\mathscr{E}}^c(\varepsilon))\right)\Big]$$

$$\leqslant \mathbb{E}\left[\sup_{y\in B_{r*}(0)\cap\hat{D}^\pm(\varepsilon^\gamma)} \mathbf{1}\{ \sup_{[T_{rec}+\kappa\gamma|\ln\varepsilon|, T/\varepsilon^\delta - s_r*]} \|X^\varepsilon(t;y) - \phi^\pm\| \geqslant h\}\mathbf{1}(\tilde{\mathscr{E}}(\varepsilon)\circ\theta_{s_r*})\right]$$

$$+ \mathbb{P}\left(\tilde{\mathscr{E}}^c(\varepsilon)\right)$$

$$\leqslant \mathbb{E}\left[\sup_{z\in B_{\varepsilon^{2\gamma}}(\phi^\pm)} \mathbf{1}\{ \sup_{t\in[0,T/\varepsilon^\delta - s_r* - T_{rec} - \kappa\gamma|\ln\varepsilon|} \|X^\varepsilon(t;y) - \phi^\pm\| \geqslant h\}\right] + \mathbb{P}\left(\tilde{\mathscr{E}}^c(\varepsilon)\right)$$

$$=: I_1 + I_2.$$

While $I_2 = I_2(\varepsilon) \to 0$ for $\varepsilon \to 0+$ by Proposition 4.11. For I_1 consider the first exit time from a small ball of radius h centered in $\phi^\pm$

$$\tau^h := \inf\{t > 0 \mid X^\varepsilon(t;x) \notin \mathscr{B}_h(\phi^\pm)\}.$$

Since it has the same exit rates as $\tau^\pm$ we get the estimate

$$I_1 \leqslant \mathbb{E}\left[\sup_{y\in B_{\varepsilon^{2\gamma}}(\phi^\pm)} \mathbf{1}\{\tau_y^h(\varepsilon) < T/\varepsilon^\delta\}\right]$$

$$\leqslant \mathbb{E}\left[\sup_{y\in B_{\varepsilon^{2\gamma}}(\phi^\pm)} \mathbf{1}\{\lambda^\pm(\varepsilon)\tau_y^h(\varepsilon) < T\lambda^\pm(\varepsilon)/\varepsilon^\delta\}\right],$$

which tends to 0 as $\varepsilon \to 0+$ by Corollary 5.9. □

Corollary 7.7 (Convergence on subcritical time scales). *Assume that Hypotheses (H.1) and (H.2) and Convention (C) are satisfied. Fix* $0 < \delta < \alpha$. *Then for all* $h > 0$ *and* $T > 0$

$$\lim_{\varepsilon\to 0+} \mathbb{E}\left[\sup_{x\in\hat{D}^\pm(\varepsilon^\gamma)} \mathbf{1}\{\|X^\varepsilon(T/\varepsilon^\delta;x) - \phi^\pm\| > h\}\right] = 0.$$

If we assume in addition that Hypothesis (H.3) is fulfilled, the process leaves the separatrix with high probability before times of the order $\frac{T}{\varepsilon^{\alpha(1-\gamma/2)}}$. Then we obtain a result of the type of Corollary 7.7 uniformly for all initial values in H and time scales including the critical time scale $\frac{T}{\lambda^0(\varepsilon)}$. Close to the separatrix we cannot decide to which domain of attraction the process tends while apart from it the previous reasoning of Theorem 7.6 continues to hold. The result is a uniform localization theorem in space.

If we include Hypothesis (H.3) into our reasoning in the situation of Theorem 7.6 we obtain a similar result. But time uniformly in the basic estimates is addressed differently. As opposed to Theorem 7.6, where the estimation is achieved uniformly on time intervals of the order $[s_{r^*}+T_{rec}+\kappa\gamma|\ln\varepsilon|, T/\varepsilon^{\delta}]$ with $\delta\in(0,\alpha)$, we are now able to treat only intervals of the shape $[T/\varepsilon^{\alpha(1-\gamma/2)}+s_{r^*}+T_{rec}+\kappa\gamma|\ln\varepsilon|, T/\lambda^0(\varepsilon)]$. In addition, by (H.3) we only control the exit times $\tau_x^0(\varepsilon)$ of the separatrix in terms of ε but we do not know into which direction this exit leads. We have no information whether $X^{\varepsilon}(\tau_x^0(\varepsilon);x)\in\hat{D}^+(\varepsilon^{\gamma})$ or $X^{\varepsilon}(\tau_x^0(\varepsilon);x)\in\hat{D}^-(\varepsilon^{\gamma})$. This is natural, since the deterministic dynamics close to the separatrix, more precisely in the vicinity of the unstable states on the separatrix, is too slow to predetermine to which domain X^{ε} tends.

Theorem 7.8. *Assume that Hypotheses (H.1), (H.2) and (H.3) and Convention (C) are satisfied and* $T_{rec},\kappa,s_{r^*}>0$ *given by Propositions 2.12 and 4.1. Set for* $T>0$ *and* $\varepsilon>0$

$$s(\varepsilon):=\frac{T}{\varepsilon^{\alpha(1-\gamma/2)}}+s_{r^*}+T_{rec}+\kappa\gamma|\ln\varepsilon|.$$

Then there is $h_0>0$ *such that for all* $T>0$ *and* $0<h\le h_0$

$$\lim_{\varepsilon\to0+}\sup_{x\in H}\mathbb{E}\left[\sup_{t\in[s(\varepsilon),\frac{T}{\lambda^0(\varepsilon)}]}\mathbf{1}\{X^{\varepsilon}(t;x)\notin B_h(\phi^+)\cup B_h(\phi^-)\}\right]=0.$$

Proof. The proof is divided into two steps. Since with a slowly varying function ℓ we have $\lambda^0(\varepsilon)=\varepsilon^{\alpha}\,\ell(1/\varepsilon)\,\mu(B_1^c(0))$, there exists $\varepsilon_0>0$ such that for $0<\varepsilon\leqslant\varepsilon_0$ we have $\frac{T}{\lambda^0(\varepsilon)}>s(\varepsilon)$. For any $x\in H$

$$\mathbb{E}\left[\sup_{t\in[s(\varepsilon),\frac{T}{\lambda^0(\varepsilon)}]}\mathbf{1}\{X^{\varepsilon}(t;x)\notin B_h(\phi^+)\cup B_h(\phi^-)\right]$$

$$\leqslant\sup_{y\in H}\mathbb{E}\left[\mathbf{1}\{X^{\varepsilon}(s(\varepsilon);y)\notin B_h(\phi^+)\cup B_h(\phi^-)\}\right].$$

We treat the cases $y\in\hat{D}^{\pm}(\varepsilon^{\gamma})$ and $y\in\hat{D}^0(\varepsilon^{\gamma})$ separately. The first case is already treated in Theorem 7.6 implying for $\delta=1-\gamma/4$

$$\lim_{\varepsilon\to0+}\mathbb{E}\left[\sup_{y\in\hat{D}^{\pm}(\varepsilon^{\gamma})}\sup_{t\in[s_{r^*}+T_{rec}+\kappa\gamma|\ln\varepsilon|,T/\varepsilon^{\delta}]}\mathbf{1}\left\{X^{\varepsilon}(t;y)\notin B_h(\phi^+)\cup B_h(\phi^-)\right\}\right]=0. \tag{7.6}$$

The second case $y \in \hat{D}^0(\varepsilon^\gamma)$ can be dealt with by the estimate

$$\sup_{y \in \hat{D}^0(\varepsilon^\gamma)} \mathbb{E}\left[\mathbf{1}\left\{X^\varepsilon(s(\varepsilon); y) \notin B_h(\phi^+) \cup B_h(\phi^-)\right\}\right]$$

$$\leqslant \sup_{y \in \hat{D}^0(\varepsilon^\gamma)} \mathbb{E}\left[\mathbf{1}\left\{X^\varepsilon(s(\varepsilon); y) \notin B_h(\phi^+) \cup B_h(\phi^-)\right\} \cap \left\{\tau_y^0(\varepsilon) \leqslant \frac{T}{\varepsilon^{\alpha(1-\gamma/2)}}\right\}\right]$$

$$+ \sup_{y \in \hat{D}^0(\varepsilon^\gamma)} \mathbb{P}\left(\tau_y^0(\varepsilon) > \frac{T}{\varepsilon^{\alpha(1-\gamma/2)}}\right) =: I_1 + I_2.$$

By Lemma 7.4, $I_2 = I_2(\varepsilon)$ tends to 0 as $\varepsilon \to 0+$. To estimate the term I_1, we use the strong Markov property of X^ε to obtain

$$I_1 = \sup_{y \in \tilde{D}^0(\varepsilon^\gamma)} \mathbb{E}\left[\mathbf{1}\left\{\tau_y^0(\varepsilon) \leqslant \frac{T}{\varepsilon^{\alpha(1-\gamma/2)}}\right\}\right.$$
$$\left.\cdot\, \mathbf{1}\left\{X^\varepsilon\left(s(\varepsilon) - \tau_y^0(\varepsilon); y\right) \circ \theta_{\tau_y^0(\varepsilon)} \notin B_h(\phi^+) \cup B_h(\phi^-)\right\}\right], \tag{7.7}$$

and consequently

$$I_1 \leqslant \sup_{y \in \hat{D}^0(\varepsilon^\gamma)} \mathbb{E}\Big[\mathbf{1}\{\tau_y^0(\varepsilon) \leqslant \frac{T}{\varepsilon^{\alpha(1-\gamma/2)}}\}$$
$$\cdot\, \mathbf{1}\left\{X^\varepsilon\left(s(\varepsilon) - \tau_y^0(\varepsilon); y\right) \circ \theta_{\tau_y^0(\varepsilon)} \notin B_h(\phi^+) \cup B_h(\phi^-)\right\}\Big]$$
$$\leqslant \sup_{y \in \hat{D}^0(\varepsilon^\gamma)} \mathbb{E}\Big[\mathbf{1}\{\tau_y^0(\varepsilon) \leqslant \frac{T}{\varepsilon^{\alpha(1-\gamma/2)}}\}$$
$$\cdot \sup_{z \in \hat{D}^+(\varepsilon^\gamma) \cup \hat{D}^-(\varepsilon^\gamma)} \sup_{t \in [s_{r*} + T_{rec} + \kappa\gamma|\ln\varepsilon|, s(\varepsilon)]} \mathbf{1}\left\{X^\varepsilon(t; z) \notin B_h(\phi^+) \cup B_h(\phi^-)\right\}\Big]$$
$$\leqslant \sum_{k=\pm} \mathbb{E}\left[\sup_{z \in \hat{D}^k(\varepsilon^\gamma)} \mathbf{1}\left\{\sup_{t \in [s_{r*} + T_{rec} + \kappa\gamma|\ln\varepsilon|, s(\varepsilon)]} \|X^\varepsilon(t; z) - \phi^k\| > h\right\}\right].$$

Again by choosing $\delta = \alpha(1 - \gamma/4)$ Theorem 7.6 ensures that I_1 tends to zero as $\varepsilon \to 0+$. □

Corollary 7.9. *Assume that Hypotheses (H.1), (H.2) and (H.3) and Convention (C) are satisfied. Fix* $\delta \in (\alpha(1 - \gamma/2), \alpha)$. *Then there is* $h_0 > 0$ *such that for all* $T > 0$ *and* $0 < h \leq h_0$

$$\lim_{\varepsilon\to 0+}\sup_{x\in H}\mathbb{E}\left[\sup_{t\in[\frac{T}{\varepsilon^{\delta}},\frac{T}{\lambda^0(\varepsilon)}]}\mathbf{1}\{X^{\varepsilon}(t;x)\notin B_h(\phi^+)\cup B_h(\phi^-)\}\right]=0.$$

The proof if obvious, since there is $\varepsilon_0 > 0$ such that for $0 < \varepsilon \leq \varepsilon_0$ we have

$$\frac{T}{\varepsilon^{\alpha(1-\gamma/2)}} + s_{r^*} + T_{rec} + \kappa\gamma|\ln\varepsilon| < \frac{T}{\varepsilon^{\delta}}.$$

7.3 Metastable Behavior

In this section, equipped with our previously obtained knowledge of exit and transition times in the limit of small noise amplitude $\varepsilon \to 0$, we shall investigate the global asymptotic behavior of our jump diffusion process in the time scale in which transitions occur, i.e. in the scale given by $\lambda^0(\varepsilon) = \nu(\frac{1}{\varepsilon}B_1^c(0)), \varepsilon > 0$. It turns out that in this time scale, the switching of the diffusion between neighborhoods of the stable solutions $\phi^{\pm}$ can be well described by a Markov chain jumping back and forth between two states with a characteristic Q-matrix determined by the quantities $\mu((D_0^{\pm})^c)/\mu(B_1^c(0))$ as jumping rates. We shall obtain convergence results for the finite-dimensional distributions and the initial values $x \in \hat{D}^{\pm}(\varepsilon^{\gamma})$. This convergence result pertains even for all $x \in H$ if the Markov chain has random initial conditions described by the random points of exit into $\hat{D}^{+}(\varepsilon^{\gamma})$ or $\hat{D}^{-}(\varepsilon^{\gamma})$ of the process $X^{\varepsilon}(\cdot;x)$ starting in $\hat{D}^{0}(\varepsilon^{\gamma})$.

For $T > 0$ we consider (finite) *partitions of* $[0,T]$ as finite families of points in $[0,T]$, with $0 < t_1 < \ldots t_n = T$, and write $|\pi| = n$. We shall denote by $\Pi[0,T]$ the collection of all finite partitions in $[0,T]$. For convenience we shall write for $h > 0$, $\pi = (t_1,\ldots,t_n) \in \Pi[0,T]$ and $\bar{v} = (v_1,\ldots,v_n) \in \{\phi^+,\phi^-\}^{|\pi|}$ and $\varepsilon > 0$

$$X^{\varepsilon}(\frac{\pi}{\lambda^0(\varepsilon)};\cdot) := (X^{\varepsilon}(\frac{t_1}{\lambda^0(\varepsilon)};\cdot),\ldots,X^{\varepsilon}(\frac{t_{|\pi|}}{\lambda^0(\varepsilon)};\cdot))$$

and

$$B_h(\bar{v}) = B_h(v_1)\times\cdots\times B_h(v_n).$$

Theorem 7.10 (Metastability). *Suppose Hypotheses (H.1), (H.2) and (H.3) and Convention (C) are satisfied and denote by* μ *the limiting measure of* ν *according to (3.44). Then there exists* $h_0 > 0$ *and a continuous time Markov chain* $(Z(t))_{t\geq 0}$ *switching between the elements of* $\{\phi^+,\phi^-\}$ *with generating matrix*

$$Q = \frac{1}{\mu(B_1^c(0))}\begin{pmatrix} -\mu\left((D_0^+)^c\right) & \mu\left((D_0^+)^c\right) \\ \mu\left((D_0^-)^c\right) & -\mu\left((D_0^-)^c\right)\end{pmatrix}.$$

with the properties: for all $T > 0$, $\pi = (t_1, \ldots, t_n) \in \Pi[0,T]$, $\bar{v} \in (v_1, \ldots, v_{|\pi|}) \in \{\phi^+, \phi^-\}^{|\pi|}$ *and* $0 < h \leqslant h_0$ *we have*

$$\lim_{\varepsilon \to 0+} \sup_{x \in \hat{D}^{\pm}(\varepsilon^{\gamma})} \left| \mathbb{P}\big(X^{\varepsilon}(\frac{\pi}{\lambda^0(\varepsilon)}; x) \in B_h(\bar{v})\big) - \mathbb{P}\big(Z(\pi, x) = \bar{v}\big) \right| = 0.$$

Proof. The proof is given in four steps.

1. Construction of an auxiliary process $\hat{Z}^{\varepsilon}$:

We fix $h_0 > 0$ such that for $0 < h \leqslant h_0$ the claims of Theorem 6.7 and Corollary 7.8 hold, and define a two state auxiliary process $\hat{Z}^{\varepsilon}$ by the sequence of stopping times and stable states marking the transitions between $B_h(\phi^+)$ and $B_h(\phi^-)$. Note that $\tau_x^0(\varepsilon) = 0$ if $x \in \hat{D}^{\pm}(\varepsilon^{\gamma})$. For $x \in D^{\pm}$, $\varepsilon > 0$ and $k \geqslant 1$ let

$$\chi^0(\varepsilon; x) := 0,$$

$$m^0(\varepsilon, x) := \phi^{\pm},$$

$$m^k(\varepsilon; x) := \phi^+ \mathbf{1}\{X^{\varepsilon}(\chi^k(\varepsilon; x); x) \in B_h(\phi^+)\} + \phi^- \mathbf{1}\{X^{\varepsilon}(\chi^k(\varepsilon; x); x) \in B_h(\phi^-)\},$$

$$\chi^k(\varepsilon; x) := \inf\{t > \chi^{k-1}(\varepsilon; x) \mid X^{\varepsilon}(t; x) \in B_h(\phi^+) \cup B_h(\phi^-) \setminus B_h(m^{k-1}(\varepsilon; x))\}.$$

Based on these quantities we define a Markovian finite state process on the critical time scale $t/\lambda^0(\varepsilon)$ by setting

$$\hat{Z}^{\varepsilon}(t; x) := \sum_{k=0}^{\infty} m^k(\varepsilon; x)\, \mathbf{1}\{\chi^k(\varepsilon; x) \leqslant t/\lambda^0(\varepsilon) < \chi^{k+1}(\varepsilon; x)\}, \quad 0 \leq t \leq T.$$

By construction, the jump times of $\hat{Z}^{\varepsilon}$ are $\lambda^0(\varepsilon)\chi^k(\varepsilon; x)$ and since there are only two stable states we obtain for any $k \geqslant 2$, $x \in B_h(\phi^{\pm})$ and $\varepsilon > 0$

$$\begin{aligned} &\mathbb{P}\left(m^{k+1}(\varepsilon; x) = \phi^{\mp} \mid m^k(\varepsilon; x) = \phi^{\pm}\right) \\ &\quad = \mathbb{P}\left(\hat{Z}^{\varepsilon}(\lambda^0(\varepsilon)\chi^{k+1}(\varepsilon); x) = \phi^{\mp} \mid \hat{Z}^{\varepsilon}(\lambda^0(\varepsilon)\chi^k(\varepsilon); x) = \phi^{\pm}\right) = 1. \end{aligned} \tag{7.8}$$

This implies that $\hat{Z}^{\varepsilon}$ is a continuous time Markov chain.

2. Construction of the limiting Markov chain:

Define a continuous time Markov process $(Z(t))_{t \geq 0}$ taking values in $\{\phi^+, \phi^-\}$ by its generating matrix

$$Q = \begin{pmatrix} -q_+ & q_+ \\ q_- & -q_- \end{pmatrix} = \frac{1}{\mu(B_1^c(0))} \begin{pmatrix} -\mu\left(\left(D_0^+\right)^c\right) & \mu\left(\left(D_0^+\right)^c\right) \\ \mu\left(\left(D_0^-\right)^c\right) & -\mu\left(\left(D_0^-\right)^c\right) \end{pmatrix}.$$

Denote its jump times and states $(\chi_k, y_k = Y_{\chi_k}), k \in \mathbb{N}$, where inter jump times are conditionally independent and exponentially distributed with

$$\mathscr{L}\left(\chi_{k+1} - \chi_k | y_k = \phi^{\pm}\right) = EXP(1/q_{\pm})$$

$$\mathbb{P}\left(y_{k+1} = \phi^{\pm} | y_k = \phi^{\mp}\right) = 1.$$

For $x \in D^{\pm}$, a partition $\pi \in \Pi[0, T]$ and $\bar{v} \in \{\phi^+, \phi^-\}^{|\pi|}$ we may estimate

$$\begin{aligned}
&\left|\mathbb{P}\big(X^{\varepsilon}(\frac{\pi}{\lambda^0(\varepsilon)}; x) \in B_h(\bar{v})\big) - \mathbb{P}_{\phi^{\pm}}\big(Z(\pi) = \bar{v}\big)\right| \\
&\leqslant \left|\mathbb{P}\big(X^{\varepsilon}(\frac{\pi}{\lambda^0(\varepsilon)}; x) \in B_h(\bar{v})\big) - \mathbb{P}_{\phi^{\pm}}\big(\hat{Z}(\pi) = \bar{v}\big)\right| \\
&+ \left|\mathbb{P}_{\phi^{\pm}}\big(\hat{Z}(\pi) = \bar{v}\big) - \mathbb{P}_{\phi^{\pm}}\big(Z(\pi) = \bar{v}\big)\right|.
\end{aligned} \tag{7.9}$$

3. Convergence of the rescaled process to the auxiliary process:

Let us estimate the first term on the right-hand side of (7.9). For this purpose we cut the events of both summands into comparable pieces. For $\varepsilon > 0$, $\pi \in \Pi[0, T]$, $\bar{v} \in \{\phi^+, \phi^-\}^{|\pi|}$ and $x \in D^{\pm}$ we have

$$\begin{aligned}
&\mathbb{P}\left(X^{\varepsilon}(\pi/\lambda^0(\varepsilon); x) \in B_h(\bar{v})\right) - \mathbb{P}\left(\hat{Z}^{\varepsilon}(\pi; x) = \bar{v}\right) \\
&= \mathbb{P}\left(X^{\varepsilon}(\pi/\lambda^0(\varepsilon); x) \in B_h(\bar{v}) \text{ and } \hat{Z}^{\varepsilon}(\pi; x) = \bar{v}\right) \\
&\quad + \mathbb{P}\big(X^{\varepsilon}(\pi/\lambda^0(\varepsilon); x) \in B_h(\bar{v}) \text{ and } \hat{Z}^{\varepsilon}(\pi; x) \in \{\phi^+, \phi^-\}^{|\pi|} \setminus \{\bar{v}\}\big) \\
&\quad - \mathbb{P}\big(\hat{Z}^{\varepsilon}(\pi; x) = \bar{v} \text{ and } X^{\varepsilon}(\pi/\lambda^0(\varepsilon); x) \in B_h(\bar{v})\big) \\
&\quad - \mathbb{P}\big(\hat{Z}^{\varepsilon}(\pi; x) = \bar{v} \text{ and } X^{\varepsilon}(\pi/\lambda^0(\varepsilon); x) \in \bigcup_{w \in \{\phi^+, \phi^-\}^{|\pi|} \setminus \{\bar{v}\}} B_h(w)\big) \\
&\quad - \mathbb{P}\big(\hat{Z}^{\varepsilon}(\pi; x) = \bar{v} \text{ and } X^{\varepsilon}(\pi/\lambda^0(\varepsilon); x) \notin \bigcup_{w \in \{\phi^+, \phi^-\}^{|\pi|}} B_h(w)\big).
\end{aligned}$$

Note that the first and the third term of the right-hand side cancel and the second and the fourth term vanish by definition of $\hat{Z}^{\varepsilon}$. Therefore we are left with the last term, which we may estimate for $\delta > \alpha(1 - \gamma/2)$ by

$$
\begin{aligned}
&|\mathbb{P}(X^{\varepsilon}(\frac{\pi}{\lambda^{0}(\varepsilon)};x) \in B_h(\bar{v}) - \mathbb{P}(\hat{Z}^{\varepsilon}(\pi;x) = \bar{v})| \\
&\leqslant \mathbb{P}\big(X^{\varepsilon}(\pi/\lambda^{0}(\varepsilon);x) \notin \bigcup_{w\in\{\phi^{+},\phi^{-}\}^{|\pi|}} B_h(w)\big) \\
&\leqslant \sup_{y\in H} \mathbb{E}\left[\sup_{t\in[\frac{T}{\varepsilon^{\delta}},\frac{T}{\lambda^{0}(\varepsilon)}]} X^{\varepsilon}(t;y) \notin B_h(\phi^{+}) \cup B_h(\phi^{-})\right].
\end{aligned}
$$

The expression in the preceding line does not depend on $\bar{v} \in \{\phi^{+},\phi^{-}\}^{|\pi|}$ or $x \in D^{\pm}$, such that we can take the supremum on the left-hand side. Hence we obtain

$$
\sup_{x\in D^{\pm}} \left|\mathbb{P}\left(X^{\varepsilon}(\pi/\lambda^{0}(\varepsilon);x) \in B_h(\bar{v})\right) - \mathbb{P}\left(\hat{Z}^{\varepsilon}(\pi;x) = \bar{v}\right)\right| \leq p_1(\varepsilon).
$$

The result $\lim_{\varepsilon\to 0+} p_1(\varepsilon) = 0$ is implied by Theorem 7.8.

4. Convergence of $\hat{Y}^{\varepsilon}$ to the Markov chain:

Let us now treat the second term on the right-hand side of (7.9). Its convergence to zero is a consequence of the weak convergence of $\hat{Z}^{\varepsilon}$ to Z, and therefore implied by the joint weak convergence of jump times and jump increments (see for example [Xia92])

$$
\mathscr{L}\left(\left((\lambda^{0}(\varepsilon)\chi^{k}(\varepsilon;x), m^{k}(\varepsilon;x)\right)_{k\geq 0})\right) \to \mathscr{L}\left(\left((\chi_k, y_k)\right)_{k\geq 0}\right), \quad \varepsilon \to 0+.
$$

This in turn follows from the joint weak convergence of the inter jump times and jump increments

$$
\mathscr{L}\left(\left((\lambda^{0}(\varepsilon)(\chi^{k+1}(\varepsilon;x) - \chi^{k}(\varepsilon;x)), m^{k}(\varepsilon;x)\right)_{k\geq 0})\right) \xrightarrow{d} \mathscr{L}\left(\left((\chi_{k+1} - \chi_k, y_k)\right)_{k\geq 0}\right) \tag{7.10}
$$

as $\varepsilon \to 0$. Since $\hat{Z}^{\varepsilon}$ is a Markov chain, to verify (7.10) it is sufficient to show individual convergence for the infinitely many components. To prove the latter we shall treat the case $k \geqslant 2$ with $x \in B_h(\phi^{\pm})$ and $k = 1$ with $x \in \hat{D}^{\pm}(\varepsilon^{\gamma})$ and $x \in \hat{D}^{0}(\varepsilon^{\gamma})$ separately. Note that for $k \geq 2$ we only have to treat initial values $x \in B_h(\phi^{\pm})$, since $X^{\varepsilon}(\chi^{1}(\varepsilon;x);x) \in B_h(\phi^{\pm})$. Together with the fact that the elements of $\left(\chi^{k+1}(\varepsilon;x) - \chi^{k}(\varepsilon;x)\right)_{k\geqslant 1}$ are independent, condition (7.10) boils down to the convergence

$$
\begin{aligned}
&\mathscr{L}(\lambda^{0}(\varepsilon)(\chi^{k+1}(\varepsilon;x) - \chi^{k}(\varepsilon;x)) \mid m^{k}(\varepsilon;x) = \phi^{\pm}) \\
&\xrightarrow{d} EXP(1/q^{\pm}) = EXP(\frac{\mu(B_1^{c}(0))}{\mu((D_0^{\pm})^{c})}) \quad \text{as } \varepsilon \to 0+.
\end{aligned} \tag{7.11}
$$

To prove this, note that the strong Markov property of X^ε implies for the law of increments of transition times for $x \in B_h(\phi^\pm)$, $\varepsilon > 0$

$$\begin{aligned}
&\mathscr{L}\big(\lambda^0(\varepsilon)\,\big(\chi^{k+1}(\varepsilon;x) - \chi^k(\varepsilon;x)\big) \mid m^k(\varepsilon;x) = \phi^\pm\big) \\
&\quad = \mathscr{L}\left(\lambda^0(\varepsilon)\,\big(\chi^{k+1}(\varepsilon;x) - \chi^k(\varepsilon;x)\big) \mid X^\varepsilon(\chi^k(\varepsilon;x);x) \in B_h(\phi^\pm)\right) \\
&\quad = \mathscr{L}\left(\lambda^0(\varepsilon)\chi^1(\varepsilon; X^\varepsilon(\chi^k(\varepsilon;x);x))\right) = \mathscr{L}\left(\lambda^0(\varepsilon)\chi^\pm(\varepsilon; X^\varepsilon(\chi^k(\varepsilon;x);x))\right).
\end{aligned} \tag{7.12}$$

In addition, the regular variation with index $-\alpha$ of $\lambda^\pm$ and λ^0 implies

$$\lim_{\varepsilon \to 0+} \frac{\lambda^0(\varepsilon)}{\lambda^\pm(\varepsilon)} = \frac{\mu(B_1^c(0)}{\mu\left((D^\pm)^c\right)}. \tag{7.13}$$

Let now $(\bar{\tau}(\varepsilon))_{\varepsilon>0}$ be the family of exponentially distributed random variables with parameter 1 according to Theorem 5.11. We therefore see that it suffices to prove that for $h > 0$

$$\mathbb{E}\left[\mathbf{1}\{|\lambda^0(\varepsilon)\chi^\pm(\varepsilon;x) - \frac{\mu(B_1^c(0)}{\mu\left((D^\pm)^c\right)}\bar{\tau}(\varepsilon)| > h\}\right] \to 0, \qquad \text{as } \varepsilon \to 0+.$$

In fact, we may write

$$\begin{aligned}
&\mathbb{E}\left[\mathbf{1}\{|\lambda^0(\varepsilon)\chi^\pm(\varepsilon;x) - \frac{\mu(B_1^c(0)}{\mu\left((D^\pm)^c\right)}\bar{\tau}(\varepsilon)| > h\}\right] \\
&\quad \leqslant \mathbb{E}\left[\sup_{x \in \hat{D}^\pm(\varepsilon^\gamma)} \mathbf{1}\{|\frac{\lambda^0(\varepsilon)}{\lambda^\pm(\varepsilon)}\lambda^\pm(\varepsilon)\chi^\pm(\varepsilon;x) - \frac{\lambda^0(\varepsilon)}{\lambda^\pm(\varepsilon)}\bar{\tau}(\varepsilon)| > h/2\}\right] \\
&\qquad + \mathbb{P}\Big(\underbrace{|\frac{\lambda^0(\varepsilon)}{\lambda^\pm(\varepsilon)} - \frac{\mu(B_1^c(0)}{\mu\left((D_0^\pm)^c\right)}|}_{\to 0,\ \varepsilon \to 0+}\ \bar{\tau}(\varepsilon) > h/2\Big)
\end{aligned} \tag{7.14}$$

By (7.13), the second term of the last estimate tends to zero. For the first one we use Theorem 6.7 to conclude. □

Theorem 7.11 (Uniform Metastability). *Let the Chafee–Infante parameter $\pi^2 < \lambda \neq (k\pi)^2$ for $k \in \mathbb{N}$ be given and denote by μ the limiting measure of ν according to (3.44). Suppose that Hypotheses (H.1), (H.2) and (H.3) and Convention (C) are satisfied. Then there exists $h_0 > 0$ and a continuous time Markov chain $(Y(t;x))_{t\geq 0}$ starting in $\phi^\pm$ if $x \in D^\pm$ and switching between the elements of $\{\phi^+, \phi^-\}$ with generating matrix*

$$Q = \frac{1}{\mu(B_1^c(0))} \begin{pmatrix} -\mu\left(\left(D_0^+\right)^c\right) & \mu\left(\left(D_0^+\right)^c\right) \\ \mu\left(\left(D_0^-\right)^c\right) & -\mu\left(\left(D_0^-\right)^c\right) \end{pmatrix}$$

and random initial condition

$$\Phi^\varepsilon(x) = \begin{cases} \phi^+ \text{ if } x \in D^+ \text{ or } X^\varepsilon(\tau_x^0(\varepsilon); x) \in D^+ \text{ if } x \in \mathscr{S}, \\ \phi^- \text{ if } x \in D^- \text{ or } X^\varepsilon(\tau_x^0(\varepsilon); x) \in D^- \text{ if } x \in \mathscr{S}, \end{cases}$$

which satisfies the following. For all $T > 0$, $\pi \in \Pi[0,T]$, $\bar{v} \in \{\phi^+, \phi^-\}^{|\pi|}$ *and* $0 < h \leqslant h_0$ *we have*

$$\lim_{\varepsilon \to 0+} \sup_{x \in H} \left| \mathbb{P}\big(X^\varepsilon(\frac{\pi}{\lambda^0(\varepsilon)}; x) \in B_h(\bar{v})\big) - \mathbb{P}\big(Y(\pi, \Phi^\varepsilon(x)) = \bar{v}\big) \right| = 0.$$

Proof. We proceed in similar steps as in the previous proof. For fixed $\pi \in \Pi[0,T]$ we write $\bar{1} := (1, \ldots, 1) \in \{1\}^{|\pi|}$.

1. Construction of an auxiliary process $\hat{Z}^\varepsilon$:

Fix h_0 and $0 < h \leqslant h_0$ as in the proof of Theorem 7.10. We define again

$$\hat{Z}^\varepsilon(t; x) := \sum_{k=0}^{\infty} m^k(\varepsilon; x)\, \mathbf{1}\{\chi^k(\varepsilon; x) \leqslant t/\lambda^0(\varepsilon) < \chi^{k+1}(\varepsilon; x)\}, \quad 0 \le t \le T,$$

with slightly modified transition times and random states

$$\chi^0(\varepsilon; x) := 0,$$

$$m^0(\varepsilon, x) := \Phi^\varepsilon(x)$$

$$\chi^1(\varepsilon, x) := \tau_x^0(\varepsilon) + \chi^\pm(\varepsilon; X^\varepsilon(\tau_x^0(\varepsilon); x)),$$

$$\chi^k(\varepsilon; x) := \inf\{t > \chi^{k-1}(\varepsilon; x) \mid X^\varepsilon(t; x) \in B_h(\phi^+) \cup B_h(\phi^-) \setminus B_h(m^{k-1}(\varepsilon; x))\}$$

$$m^k(\varepsilon; x) := \phi^+ \mathbf{1}\{X^\varepsilon(\chi^k(\varepsilon; x); x) \in B_h(\phi^+)\} + \phi^- \mathbf{1}\{X^\varepsilon(\chi^k(\varepsilon; x); x) \in B_h(\phi^-)\}.$$

2. Construction of the limiting Markov chain:

The continuous time Markov process $(Z(t; x))_{t \geqslant 0}$ taking values in $\{\phi^+, \phi^-\}$ is defined identically by its generating matrix

$$Q = \begin{pmatrix} -q_+ & q_+ \\ q_- & -q_- \end{pmatrix} = \frac{1}{\mu(B_1^c(0))} \begin{pmatrix} -\mu\left(\left(D_0^+\right)^c\right) & \mu\left(\left(D_0^+\right)^c\right) \\ \mu\left(\left(D_0^-\right)^c\right) & -\mu\left(\left(D_0^-\right)^c\right) \end{pmatrix}.$$

Denote its jump times and states $(\chi_k, y_k = Z_{\chi_k}), k \in \mathbb{N}$, where inter jump times are conditionally independent and exponentially distributed with

$$\mathscr{L}\left(\chi_{k+1} - \chi_k | y_k = \phi^{\pm}\right) = EXP(1/q_{\pm})$$

$$\mathbb{P}\left(y_{k+1} = \phi^{\pm} | y_k = \phi^{\mp}\right) = 1.$$

For $\varepsilon > 0$, $x \in D^{\pm}$, a partition $\pi \in \Pi[0, T]$ and $\bar{v} \in \{\phi^+, \phi^-\}^{|\pi|}$ we may estimate

$$\begin{aligned}
&\left|\mathbb{P}\big(X^{\varepsilon}(\frac{\pi}{\lambda^0(\varepsilon)}; x) \in B_h(\bar{v})\big) - \mathbb{P}\big(Z(\pi; \Phi^{\varepsilon}(x)) = \bar{v}\big)\right| \\
&\quad \leqslant \sup_{x \in \hat{D}^{\pm}(\varepsilon^{\gamma})} \left|\mathbb{P}\big(X^{\varepsilon}(\frac{\pi}{\lambda^0(\varepsilon)}; x) \in B_h(\bar{v})\big) - \mathbb{P}\big(Z(\pi; \Phi^{\varepsilon}(x)) = \bar{v}\big)\right| \\
&\quad + \sup_{x \in \hat{D}^0(\varepsilon^{\gamma})} \left|\mathbb{P}\big(X^{\varepsilon}(\frac{\pi}{\lambda^0(\varepsilon)}; x) \in B_h(\bar{v})\big) - \mathbb{P}\big(\hat{Z}(\pi; X^{\varepsilon}(\tau_x^0(\varepsilon); x)) = \bar{v}\big)\right| \\
&\quad + \sup_{x \in \hat{D}^0(\varepsilon^{\gamma})} \left|\mathbb{P}\big(\hat{Z}(\pi; X^{\varepsilon}(\tau_x^{\pm}(\varepsilon); x)) = \bar{v}\big) - \mathbb{P}\big(Z(\pi; \Phi^{\varepsilon}(x)) = \bar{v}\big)\right| = I_1 + I_2 + I_3.
\end{aligned} \tag{7.15}$$

The convergence of I_1 to 0 as $\varepsilon \to 0+$ is covered by Theorem 7.10.

3. Convergence of the rescaled process to the auxiliary process:

To show the convergence of I_2 for $\pi = (t_1, \ldots, t_{|\pi|})$ we first choose $\varepsilon > 0$ small enough to have $\frac{t_1}{\lambda^0(\varepsilon)} > \frac{T}{\varepsilon^{\alpha(1-\gamma/2)}}$ and then use the strong Markov property at $\tau_x^0(\varepsilon)$ to obtain

$$\begin{aligned}
I_2 &\leqslant \sup_{x \in \hat{D}^0(\varepsilon^{\gamma})} \left|\mathbb{P}\big(X^{\varepsilon}(\frac{\pi}{\lambda^0(\varepsilon)}; x) \in B_h(\bar{v})\big) - \mathbb{P}\big(\hat{Z}(\pi; X^{\varepsilon}(\tau_x^0(\varepsilon); x)) = \bar{v}\big)\right| \\
&\leqslant \sup_{x \in \hat{D}^0(\varepsilon^{\gamma})} \Big|\mathbb{P}\big(X^{\varepsilon}(\frac{\pi}{\lambda^0(\varepsilon)} - \tau_x^0(\varepsilon)\bar{1};\ X^{\varepsilon}(\tau_x^0(\varepsilon)\bar{1}; x)) \in B_h(\bar{v}),\ \tau_x^0(\varepsilon) \leqslant \frac{T}{\varepsilon^{\alpha(1-\gamma/2}}\big) \\
&\qquad - \mathbb{P}\big(\hat{Z}(\pi; X^{\varepsilon}(\tau_x^0(\varepsilon); x)) = \bar{v}\big)\Big| + \sup_{x \in \hat{D}^0(\varepsilon^{\gamma})} \mathbb{P}(\tau_x^0(\varepsilon) > \frac{T}{\varepsilon^{\alpha(1-\gamma/2)}}) \\
&\leqslant \sup_{y \in \hat{D}^+(\varepsilon^{\gamma}) \cup \hat{D}^-(\varepsilon^{\gamma})} \left|\mathbb{P}\big(X^{\varepsilon}(\frac{\pi}{\lambda^0(\varepsilon)} - \frac{T}{\varepsilon^{\alpha(1-\gamma/2)}}\bar{1}; y) \in B_h(\bar{v})\big) - \mathbb{P}\big(\hat{Z}(\pi; y) = \bar{v}\big)\right| \\
&\quad + \sup_{x \in \hat{D}^0(\varepsilon^{\gamma})} \mathbb{P}(\tau_x^0(\varepsilon) > \frac{T}{\varepsilon^{\alpha(1-\gamma/2)}}) = I_4 + I_5.
\end{aligned} \tag{7.16}$$

Now $I_5 \to 0$ as $\varepsilon \to 0+$ by Lemma 7.4 under Hypothesis (H.3). Since

$$|X^\varepsilon(\frac{\pi}{\lambda^0(\varepsilon)} - \frac{T}{\varepsilon^{\alpha(1-\gamma/2)}}\bar{1}; y) - X^\varepsilon(\frac{\pi}{\lambda^0(\varepsilon)}; y)| \xrightarrow{\mathbb{P}} 0 \text{ as } \varepsilon \to 0+,$$

the same convergence holds in law.

4. Convergence of the auxiliary process to the Markov chain:

We argue in the same way as in Part 4 of the proof of Theorem 7.10 remarking that the convergence of I_3 to 0 as $\varepsilon \to 0+$ in (7.15) is a consequence of

$$\mathscr{L}\left(((\lambda^0(\varepsilon)(\chi^{k+1}(\varepsilon;x) - \chi^k(\varepsilon;x)), m^k(\varepsilon;x))_{k\geq 0})\right) \xrightarrow{d} \mathscr{L}\left(((\chi_{k+1} - \chi_k, y_k))_{k\geq 0}\right). \tag{7.17}$$

Let now $\bar{\tau}$ be exponentially distributed with parameter 1 according to Theorem 5.11. Recalling that it is sufficient to establish the individual convergence of the components, we remark that the case $k \geqslant 2$ has already been taken care of in (7.14). For $k = 1$ and $x \in H$ the convergence is a consequence of

$$\begin{aligned}
&\mathbb{E}\left[\mathbf{1}\{|\lambda^0(\varepsilon)(\chi^\pm(\varepsilon;x) + \tau_x^0(\varepsilon)) - \frac{\mu(B_1^c(0)}{\mu\left((D^\pm)^c\right)}\bar{\tau}| > h\}\right] \\
&\leqslant \sup_{y\in H} \mathbb{E}\left[\mathbf{1}\{|\frac{\lambda^0(\varepsilon)}{\lambda^\pm(\varepsilon)}\lambda^\pm(\varepsilon)(\chi^\pm(\varepsilon;y) + \tau_y^0(\varepsilon)) - \frac{\lambda^0(\varepsilon)}{\lambda^\pm(\varepsilon)}\bar{\tau}| > h/2,\right. \\
&\qquad \left.\tau_y^0(\varepsilon) \leq \frac{T}{\varepsilon^{\alpha(1-\gamma/2)}}\}\right] \\
&\quad + \sup_{y\in H} \mathbb{P}(\tau_y^0(\varepsilon) > \frac{T}{\varepsilon^{\alpha(1-\gamma/2)}}) + \mathbb{P}\Big(\underbrace{|\frac{\lambda^0(\varepsilon)}{\lambda^\pm(\varepsilon)} - \frac{\mu(B_1^c(0)}{\mu\left((D_0^\pm)^c\right)}|}_{\to 0,\ \varepsilon\to 0+}\ \bar{\tau} > h/2\Big).
\end{aligned} \tag{7.18}$$

By (7.13), the third term on the right hand side of the last estimate tends to zero. By Lemma 7.4 the second summand does. For the first one we use Theorem 6.7 to conclude. □

Appendix A
The Source of Stochastic Models in Conceptual Climate Dynamics

The variability of global climate patterns for the last decades has received overwhelming interest recently. The impact human activities might have on the current terrestrial climate balance underlines the need for reliable climate modeling and simulation. The mathematical models underlying modern simulations are very complex and high dimensional. The closer to reality the resulting virtual pictures are, the closer our understanding of their contents is to our understanding of real climate. This possibly just means that it may be equally poor. In addition, climatology is a science without experiments or empirical inference in the usual sense, apart from the reproduction of past climate patterns by statistical inference from paleo-climatic data. The cross-validation of simulation output with these data is usually rather difficult. As a consequence, there certainly is the danger of too much confidence in the simulation output from the models, and the virtual world they create. And it is certainly wrong to consider computer experiments as acceptable compensation for lack of real experiments and empirical data. Therefore a physical or analytical understanding of the phenomena both in the real as in the virtual world of model simulations through conceptual insight is of central importance. It can be provided by considering conceptual, analytically accessible stochastic reductions of the complex models. Accordingly, stochastic model reduction in climate dynamics is of paramount importance.

A.1 A Conceptual Approach to Low-Dimensional Climate Dynamics

One of the main obstacles of climate modeling is the substantial variability on spatial and temporal scales ranging over many orders of magnitude. It reaches from turbulent eddies in the ocean surface due to breaking waves, through mid-latitude cyclonic storms hundreds of kilometers in extent and lasting for days, to millennial scale shifts in ice cover and ocean circulation. The low-lying physical

A. Debussche et al., *The Dynamics of Nonlinear Reaction-Diffusion Equations with Small Lévy Noise*, Lecture Notes in Mathematics 2085, DOI 10.1007/978-3-319-00828-8,

description behind imposes important mutual dependencies of quantities on these highly different time scales, which in general cannot be resolved entirely. This spread poses major challenges for any quantitatively accurate and computationally feasible representation.

To account for this variety of effects on very different scales, climatologists developed a big collection of models which are commonly classified into three groups. On the top level of quantitative accuracy are the comprehensive General Circulation Models (GCMs). These are the quantitatively most ambitious models, which attempt to represent the climate system in as much detail as computational resources and conceptional reasoning allow. Earth System Models of Intermediate Complexity (EMICs) instead are models of a more restrained resolution, which attempt to represent some subsystem of Earth's climate in detail, such as the ocean, the land surface or the atmosphere, while the interaction with other subsystems as well as external forcing remains parameterized. At the bottom of the model hierarchy according to [CMW$^+$02] are low dimensional ones such as for instance energy balance models, that ignore almost all quantities and their interactions, except for a few. They are studied under highly idealized conditions, such that they are hardly of quantitative relevance. Their interest lies in their accessibility for mathematical analysis. Very often they are completely solvable and entirely understandable. They may qualitatively predict phenomena encountered in more complex models. Their reduced complexity can help to develop conceptual qualitative paradigms capable to interpret and understand simulations obtained on the basis of EMICs or GCMs. Classical examples of this are the prediction of multiple states of the thermohaline circulation by Stommel [Sto61], of the phenomenon of sensitivity to initial conditions by Lorenz [Lor63], and of glacial metastability.

In the lower levels of climate modeling it is crucial to decide which processes to represent explicitly, which to parameterize, and how to justify or even construct the parametrization. Following [IM02], in an updated version of the traditional approach an analogy with thermodynamic limit theorems is used: by taking the proportion of scales to an infinite limit, a complete separation of micro and macro scales is obtained. In a first step, averaging of small scale processes produces deterministic dynamics for the large scale processes. In a second step, the fluctuation of the large scale variables around the averaged values of the small scale quantities is expressed by stochastic differential or partial differential equations, in which the large scale variables are driven by random processes representing the small scale components. The mathematically rigorous derivation of such equations by Khasminskii [Kha66] leads to *linear systems*, however.

A.1.1 Hasselmann's Unfinished Program

There have been serious attempts to derive simple *non-linear* climate models with stochastic forcing from idealized GCMs. This project is labeled "Hasselmann's program" after an article by Arnold [Arn01], in which the ideas by the climatologist [Has76] dating back to the mid-seventies are translated into modern mathematical language. Hasselmann's work is explicitly aimed at increasing the mathematical and physical understanding of more resolved climate models.

We shall briefly sketch the main ideas. In a first step an idealized GCM is considered as a large system of coupled ordinary (or partial) differential equations, in which for $0 < \varepsilon \ll 1$ the climate state $z = (x^\varepsilon, y^\varepsilon)$ can be separated into "slow" $x^\varepsilon(t, y^\varepsilon)$ and "fast" variables $y^\varepsilon(\frac{1}{\varepsilon}t, x^\varepsilon)$. Such a system can be formally described by

$$\begin{aligned}\dot{x}^\varepsilon &= f(x^\varepsilon, y^\varepsilon),\\ \dot{y}^\varepsilon &= \frac{1}{\varepsilon}g(x^\varepsilon, y^\varepsilon).\end{aligned}$$

The scale separation should be described by a small parameter ε corresponding to the "response time" of the scales of slow and fast variables. Now define in physical jargon $u^\varepsilon(t) := \langle x^\varepsilon(t, \cdot)\rangle, t \geqslant 0$, as an "average" of the slow variables with respect to an invariant measure of the subsystem of the fast ones. This should lead to an averaged ordinary or partial differential equation

$$\dot{u}^\varepsilon = F(u^\varepsilon),$$

where $F(u^\varepsilon) := \langle f(x^\varepsilon, \cdot)\rangle$. The first mathematically rigorous proof of such a procedure was given by Bogolyubov and Mitropolskii [BM61], establishing that under appropriate assumptions $\lim_{\varepsilon\to 0+} x^\varepsilon(t) = u^0(t)$.

In a second step, the fluctuation $x^\varepsilon(t) - u^0(t)$ of the solution around the averaged one is studied. Khasminskii [Kha66] discovered that for $t \in [0, T]$

$$L^\varepsilon(t) = \frac{1}{\sqrt{\varepsilon}}\left(x^\varepsilon(t) - u^0(t)\right)$$

has a limiting Gaussian law as $\varepsilon \to 0+$. This way, he obtains linear differential equations for the slow variables with a stochastic term replacing the fast ones on finite intervals. In the framework of diffusion limits, deviations from averaged behavior produce non-linear (partial) differential equations with stochastic forcing (see [AK01] and [MTE99]). In this reduction, an assumption is crucial that is usually very hard to rigorously establish: mixing properties of the fast components, which lead to a decay of correlations viewed by an equilibrium measure. Even in simple ocean models studied in [Maa94] coupled to a Lorenz equation as atmospheric component, different regimes of the fast motion that are only partially chaotic, complicate the mathematical treatment.

Yet many qualitative phenomena could not be captured by these methods, since they happen on ε-dependent time scales, that tend to be large for small ε, i.e. on intervals $[0, T(\varepsilon)]$, where $T(\varepsilon) \to \infty, \varepsilon \to 0+$. Among these are for example the Markovian transitions between stable states of the deterministic system that become metastable by the action of noise.

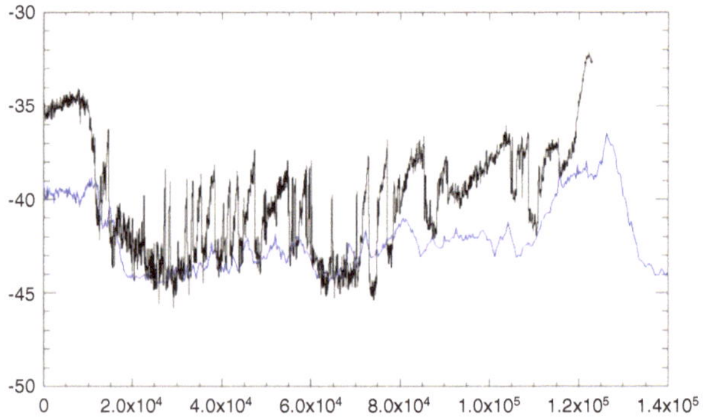

Fig. A.1 Greenland ice core $\delta^{18}O$ temperature proxies ([NGR04] core data, *black line*), 50 year average, from 120,000 years before present until now. The higher the values the warmer the average temperature [Public domain figure]

The systematic mathematical deduction of these stochastically forced equations from deterministic models remains a challenge some 35 years after their heuristic derivation by Hasselmann.

A.1.2 Energy Balance Models Perturbed by Noise of Small Intensity

An alternative approach for obtaining relevant conceptual models in climate dynamics short-circuits the derivation according to Hasselmann's program. It consists in the explicit study of given paleoclimatic time series, and the selection of the best fitting dynamical model through statistical inference. Assume that the data in the time series are realized by one of a family of deterministic dynamical systems perturbed by additive stochastic noise. Assume further that the noise is parameterized by a parameter located in a set in Euclidean space. To choose the best fitting one among the dynamical models, one has to develop a statistical test for instance for the noise parameter—often a rather hard task.

For a paleoclimatic time series from the Greenland ice shelf (Fig. A.1) providing proxies for the yearly average temperatures of the last glacial period, climatologists around [Dit99] proposed an energy balance model perturbed by heavy-tailed α-stable noise of small intensity. A statistical analysis on a physical level of rigor was used to estimate the best fitting α.

Recently this conclusion has been supported strongly by mathematical studies. In [HIP09, GHIP11] the model selection problem for the Greenland temperature time series was carried out successfully. The class of models considered is given by a dynamical system driven by a one dimensional additive α-stable process. Based on a path-wise roughness analysis using the power variations of trajectories an estimator

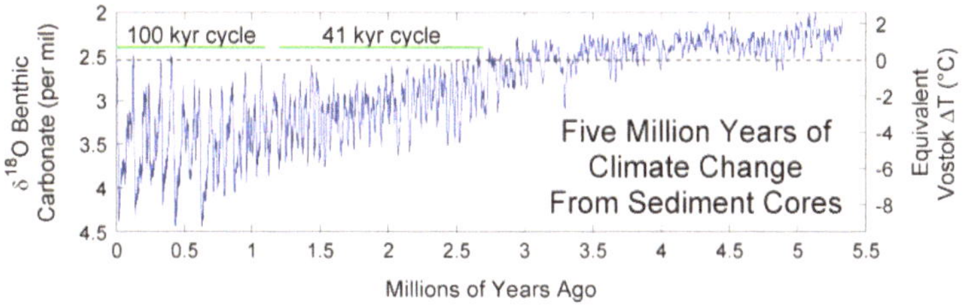

Fig. A.2 Temperature proxy for the last five million years [LR05,PJR+99] [Public domain figure]

for $\alpha \in (0, 2)$ is established. The convergence quality of this method to a unique parameter gives at least a good indication that such a signal is observed in the time series.

A.1.3 The Motivating Phenomenon: Paleoclimatic Warming Events

In the literature the term "ice age" has different meanings. In this part we adopt the following convention. *Ice age* denotes a period of lower temperature of Earth's surface and atmosphere on a scale corresponding roughly to Earth's age, i.e. on a billion to hundred million year scale. During an ice age, frequent expansions and retreats of continental ice sheets, polar ice caps and alpine glaciers are observed. These episodes of extra cold climate are called *glacial periods*. See [IR10].

Since 2.58 million years before present polar ice shields appear to reemerge, resulting in the current Quaternary Ice Age, during which around 47 glacial periods have taken place so far (See Fig. A.2).

The eventual causes for the onset of an ice age are not very clear yet. Instead, the succession of glaciation periods at least during the current ice age is closely linked to the periodic behavior of some of Earth's orbital parameters, the so-called Milankovich cycles.

The theory of climate variability due to the change in planetary orbital parameters goes back to the Serbian civil engineer M. Milankovich (1879–1958). In collaboration with W. Köppen, a German meteorologist, he recognized that the decrease of summer insolation at high latitudes may be responsible for the growth of glaciers. He expresses Earth's incoming solar radiation at a given point on the surface and time as a function of the orbital parameters, but is unsure about the critical latitude to trigger a glaciation period.

If we suppose that Earth's orbit around the sun lies approximately in a plane, it can be decomposed into three major components. The *eccentricity* of the elliptic annual trajectories of Earth around the sun vary regularly over time with periodic components of about 100,000 years. Earth's axis of rotation has an *inclination* with respect to the normal of the orbital plane, the angle of which varies between

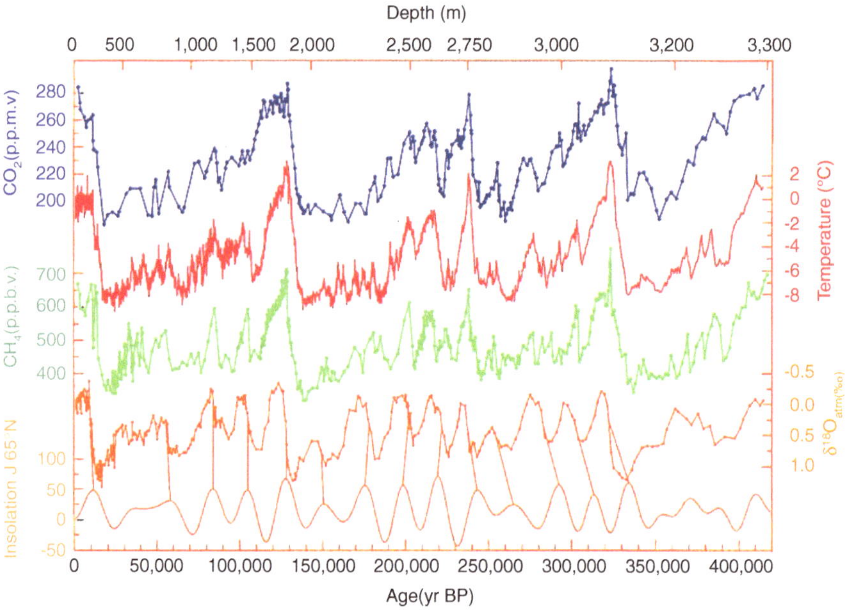

Fig. A.3 420,000 years of ice core data from [PBL$^+$97, PJR$^+$99], Antarctica research station, 6 from *bottom* to *top*: Solar variation at 65° due to Milankovich cycles (connected to $1^{18}O$), ^{18}O isotope of oxygen, levels of methane, relative temperature [Public domain figure]

22.1 and 24.5° with an approximate period of 41,000 years. It influences the solar radiation influx at high latitudes, see [Har94]. A third component is contributed by the periodic *precession* of the equinoxes, i.e. the gyration of Earth's rotation axis around the normal of the orbital plane with major periods of 19,000 and 23,000 years.

The combined effect of these three components accounts for up to 30 % of incoming solar radiation at high latitudes. The diagram of Fig. A.3 exhibits a fairly good correspondence of the summer insolation at 65° North calculated on the basis of this orbital forcing.

In the long-range data plot in Fig. A.2 one recognizes the dominant periodicity of 41,000 years until one million years ago which is replaced by the 100,000 year periodicity since then. For a recent discussion of this phenomenon see [Dit09].

The present work is motivated by a phenomenon observed during the last glacial period, about 100,000–10,000 years before present. Temperature proxies in the Greenland ice core indicate that the orbital forcing discussed above does not have a major effect within this period, and temperatures do not stay uniformly low. Instead one can recognize at least 21 major spikes, indicating abrupt extraordinary local increases by up to 8° within less than 30 years, followed by a gradual decline during several centuries (see [IR10]). The distribution of the spikes in Fig. A.1 is rather regular over the whole period.

The origin of these patterns is not quite clear. In the literature the spikes are classified into two categories. The first one consists of so-called Heinrich events. They are thought to be caused by ice sheet instabilities with a huge discharge of icebergs, i.e. enormous fresh water influx into the Atlantic. Between three and six rapid coolings are considered to be of Heinrich type. The remaining spikes are named Dansgaard–Oeschger events after their discoverers. There is so far no good explanation for their emergence. Some authors, for instance [GR01, Rah03] and [DKA06], suggest a superposition of short periodic signals of solar radiation, leading to temperature evolutions with periodic intervals that determine the Dansgaard–Oeschger events. If in this case the system retains several stable states separated by temperature thresholds those may be overcome by random perturbations. Such a phenomenon ist often referred to as *stochastic resonance*.

References

[AG79] A. Araujo, A. Giné,On tails and domains of attraction of stable measures in Banach spaces. Trans. Am. Math. Soc. **1**, 105–119 (1979)

[AK01] L. Arnold, Y. Kifer, Nonlinear diffusion approximation of the slow motion in multiscale systems (2001). Preprint

[Arn01] L. Arnold, Hasselmann's program revisited: The analysis of stochasticity in deterministic climate models, in *Stochastic Climate Models*, ed. by P. Imkeller, J.-S. von Storch. Progress in Probability (Birkhauser, Basel, 2001)

[AW00] D. Applebaum, J.-L. Wu, Stochastic partial differential equations driven by Lévy space-time white noise. Random Oper. Stoch. Equat. **3**, 245–261 (2000)

[AWZ98] S. Albeverio, J.-L. Wu, T.-Sh. Zhang, Parabolic SPDEs driven by Poisson white noise. Stoch. Process. Appl. **74**(1), 21–36 (1998)

[BEGK04] A. Bovier, M. Eckhoff, V. Gayrard, M. Klein, Metastability in reversible diffusion processes I: Sharp asymptotics and exit times. J. Eur. Math. Soc. **6**(4), 399–424 (2004)

[BGT87] N.H. Bingham, C.M. Goldie, J.L. Teugels, *Regular Variation* (Cambridge University Press, Cambridge, 1987)

[Bie98] E. Saint Loubert Bie, Etude d'une EDPS conduite par un bruit Poissonien. Probab. Theor. Relat. Fields **111**, 287–321 (1998)

[BM61] N. Bogolyubov, Y. Mitropolskii, *Asymptotic Methods in the Theory of Non-linear Oscillations* (Gordon and Breach, New York, 1961)

[Bra91] S. Brassesco, Some results on small random perturbations of an infinite-dimensional dynamical system. Stoch. Process. Appl. **38**, 33–53 (1991)

[Bre83] H. Brezis, *Functional Analysis* (Masson, Paris, 1983)

[BW06] L. Bo, Y. Wang, Stochastic Cahn-Hilliard partial differential equations with Lévy space-time white noise. Stoch. Dynam. **6**, 229–244 (2006)

[CF11] S. Cerrai, M. Freidlin, Approximation of quasi-potentials and exit problems for multidimensional RDEs with noise. Trans. Am. Math. Soc. **363**(7), 3853–3892 (2011)

[CFO11] M. Caruana, P. Friz, H. Oberhauser, A (rough) pathwise approach to a class of nonlinear SPDEs. Ann. l'Institut Henri Poincaré/Analyse non linéaire **28**, 27–46 (2011)

[CH98] T. Cazenave, A. Haraux, in *An Introduction to Semilinear Evolution Equations*. Oxford Lecture Series in Mathematics and Its Applications, vol. 13 (Oxford Science Publications, Oxford, 1998)

[Cho07] P.L. Chow, in *Stochastic Partial Differential Equations*. Applied Mathematics and Nonlinear Science Series (Chapman and Hall/CRC, New York, 2007)

[Chu01] I. Chueshov, Order-preserving skew product flows and nonautonomous parabolic systems. Acta Appl. Math. **66**(1), 185–205 (2001)

A. Debussche et al., *The Dynamics of Nonlinear Reaction-Diffusion Equations with Small Lévy Noise*, Lecture Notes in Mathematics 2085, DOI 10.1007/978-3-319-00828-8,

[Chu02] I.D. Chueshov, *Introduction to the Theory of Infinite-Dimensional Dynamical Systems* (Acta Scientific Publishing House, Kharkov, 2002)

[CI74] N. Chafee, E.F. Infante, A bifurcation problem for a nonlinear partial differential equation of parabolic type. Appl. Anal. **4**, 17–37 (1974)

[CM87] A. Chojnowska-Michalik, On processes of Ornstein-Uhlenbeck type in Hilbert spaces. Stochastics **21**, 251–286 (1987)

[CMW$^+$02] M. Claussen, L.A. Mysak, A.J. Weaver, M. Crucix, T. Fichefet, M.-F. Loutre, S.L. Weber, J. Alcamo, V.A. Alexeev, A. Berger, R. Calov, A. Ganopolski, H. Goosse, G. Lohmann, F. Lunkeit, I.I. Mokhov, V. Petoukhov, P. Stone, Z. Wang, Earth system models of intermediate c1omplexity: Closing the gap in the spectrum of climate system models. Clim. Dynam. **18**, 579–586 (2002)

[CP89] J. Carr, R. Pego, Metastable patterns in solutions of $u_t = u_{xx} - f(u)$. Comm. Pure. Appl. Math. **42**(5), 523–574 (1989)

[Dit99] P.D. Ditlevsen, Observation of a stable noise induced millennial climate changes from an ice-core record. Geophys. Res. Lett. **26**(10), 1441–1444 (1999)

[Dit09] P.D. Ditlevsen, Bifurcation structure and noise-assisted transitions in the Pleistocene glacial cycles. Paleoceanography **24**, 3204–3215 (2009)

[DKA06] P.D. Ditlevsen, M.S. Kristensen, K.K. Andersen, The recurrence time of Dansgaard-Oeschger events and limits to the possible periodic forcing. J. Clim. **18**, 2594–2603 (2006)

[DX10] Z. Dong, Y. Xie, Ergodicity of linear SPDE driven by Lévy noise. J. Syst. Sci. Complex **23**(1), 137–152 (2010)

[DZ92] G. DaPrato, J. Zabczyk, in *Stochastic Equations in Infinite Dimensions*. Encyclopedia of Mathematics and Its Applications, vol. 44 (Cambridge University Press, Cambridge, 1992)

[EFNT94] A. Eden, C. Foias, B. Nicolaenko, R. Temam, *Exponential Attractors for Dissipative Evolution Equations* (Wiley/Massons, New York/Paris, 1994)

[FJL82] W.G. Faris, G. Jona-Lasinio, Large fluctuations for a nonlinear heat equation with noise. J. Phys. A **10**, 3025–3055 (1982)

[Fou00] N. Fournier, Malliavin calculus for parabolic SPDEs with jumps. Stoch. Process. Appl. **87**, 115–147 (2000)

[Fou01] N. Fournier, Support theorem for solutions of a white noise driven SPDEs with temporal Poissonian jumps. Bernoulli **7**, 165–190 (2001)

[FR00] M. Fuhrmann, M. Röckner, Generalized Mehler semigroups: the non-Gaussian case. Potential Anal. **12**, 1–47 (2000)

[Fre85] M.I. Freidlin, Limit theorems for large deviations and reaction-diffusion equations. Ann. Probab. **13**(3), 639–675 (1985)

[FS82] B. Fiedler, A. Scheel, Spatio-temporal dynamics of reaction-diffusion patterns. J. Phys. A Math. Gen. **15**, 3025–3055 (1982)

[FTT10a] D. Filipović, S. Tappe, J. Teichmann, Jump-diffusions in Hilbert Spaces: Existence, stability and numerics. Stochastics **82**(5), 475–520 (2010)

[FTT10b] D. Filipović, S. Tappe, J. Teichmann, Term structure models driven by Wiener processes and Poisson measures: Existence and positivity. SIAM J. Financ. Math. **1**, 523–554 (2010)

[FV70] M.I. Freidlin, A.D. Ventsell, On small random perturbations of dynamical systems (English). Russ. Math. Surv. **25**(1), 1–55 (1970)

[FV98] M.I. Freidlin, A.D. Ventsell, in *Random Perturbations of Dynamical Systems*, transl. from the Russian by J. Szuecs, 2nd edn. Grundlehren der Mathematischen Wissenschaften, vol. 260 (Springer, New York, 1998)

[GHIP11] J. Gairing, C. Hein, P. Imkeller, I. Pavlyukevich, Dynamical models of climate time series and the rate of convergence of power variations. *Mathematisches Forschungsinstitut Oberwolfach, Report*, no. 40, 2011

[GR01] L. Ganopolski, S. Rahmstorf, Rapid changes of glacial climate simulated in a coupled climate model. Nature **409**, 153 (2001)

[Hai11] M. Hairer, Rough stochastic PDEs. Comm. Pure Appl. Math. **64**(11), 1547–1585 (2011)

[Hai13] M. Hairer, Solving the KPZ equation. Ann. Math. (2013) (to appear)

[Hal83] J.K. Hale, in *Infinite-Dimensional Dynamical Systems. Geometric Dynamics* (Rio de Janeiro, 1981). Lecture Notes in Mathematics, vol. 1007 (Springer, Berlin, 1983)
[Har94] D.L. Hartmann, *Global Physical Climatology*. International Geophysics Series, vol. 56 (Academic, San Diego, 1994)
[Has76] K. Hasselmann, Stochastic climate models: Part I. Theory. Tellus **28**, 473–485 (1976)
[Hau05] E. Hausenblas, Existence, uniqueness and regularity of parabolic SPDEs driven by Poisson random measure. Electron. J. Probab. **10**, 1496–1546 (electronic) (2005)
[Hau06] E. Hausenblas, SPDEs driven by Poisson random measure with non-Lipschitz coefficients: Existence results. Probab. Theor. Relat. Fields **137**, 161–200 (2006)
[Hen83] D. Henry, *Geometric Theory of Semilinear Parabolic Equations* (Springer, Berlin, 1983)
[Hen85] D. Henry, Some infinite-dimensional Morse-Smale systems defined by parabolic partial differential equations. J. Differ. Equat. **59**, 165–205 (1985)
[HIP09] C. Hein, P. Imkeller, I. Pavlyukevich, Limit theorems for p-variations of solutions of SDEs driven by additive stable Lévy noise and model selection for paleo-climatic data. Interdiscip. Math. Sci. **8**, 137–150 (2009)
[HL06] H. Hulk, F. Lindskog, Regular variation for measures on metric spaces. Publi. l'Institut de Mathématiques Nouvelle Sér. **80**, 94 (2006)
[Högl1] M. Högele, Metastability of the Chafee-Infante equation with small heavy-tailed Lévy noise - a conceptual climate model. Ph.D. thesis, Humboldt-Universität zu Berlin, 2011
[HRW12] M. Hairer, M.D. Ryser, H. Weber, Triviality of the 2d stochastic Allen-Cahn equation. Electron. J. Probab. **17**(39), 14 (2012)
[IM02] P. Imkeller, A. Monahan, Conceptual stochastic climate models. Stoch. Dynam. **2**, 311–326 (2002)
[Imk01] P. Imkeller, Energy balance models: Viewed from stochastic dynamics. Prog. Probab. **49**, 213–240 (2001)
[IP06a] P. Imkeller, I. Pavlyukevich, First exit times of SDEs driven by stable Lévy processes. Stoch. Process. Appl. **116**(4), 611–642 (2006)
[IP06b] P. Imkeller, I. Pavlyukevich, Lévy flights: Transitions and meta-stability. J. Phys. A Math. Gen. **39**, L237–L246 (2006)
[IP08] P. Imkeller, I. Pavlyukevich, Metastable behaviour of small noise Lévy-driven diffusions. ESAIM Probab. Stat. **12**, 412–437 (2008)
[IR10] Working Group 2 IPCC-Report. Glossary working group 2. Fourth Assesment Report of the Intergouvernmental Panel on Climate Change (2010), www.ipcc.ch/pdf/glossary/ar4-wg2.pdf
[Kal97] O. Kallenberg, *Foundations of Modern Probability*, 2nd edn. (2002) (Springer, New York, 1997)
[Kha66] R.Z. Khasminskii, A limit theorem for solutions of differential equations with random right-hand side. Theor. Probab. Appl. **11**, 390–406 (1966)
[Kle05] A. Klenke, *Wahrscheinlichkeitstheorie, vol. 2. korrigierte Auflage* (2008) (Springer, New York, 2005)
[Kno04] C. Knoche, SPDEs in infinite dimension with Poisson noise. C. R. Acad. Sci. Paris Ser. I **339**, 647–652 (2004)
[Kot08] P. Kotelenez, *Stochastic Ordinary and Stochastic Partial Differential Equations, Transitions from Microscopic to Macroscopic Dynamics* (Springer, New York, 2008)
[KPA88] G. Kallianpur, V. Perez-Abreu, Stochastic evolution equations driven by nuclear space valued martingales. Appl. Math. Optim. **3**, 237–272 (1988)
[KR07] N.V. Krylov, B.L. Rozovskii, Stochastic evolution equations, in *Stochastic Differential Equations: Theory and Applications*. A Volume in Honor of Prof. Boris L. Rozovskii, ed. by P. Baxendale et al. Interdisciplinary Mathematical Sciences, vol. 2 (World Scientific, Hackensack, 2007), pp. 1–69
[KRAD+08] D. Khoshnevisan, F. Rassoul-Agha, R. Dalang, C. Mueller, D. Nualart, Y. Xiao, in *A minicourse on SPDE*. Lecture Notes in Mathematics (Springer, New York, 2008)
[Lor63] E.N. Lorenz, Deterministic nonperiodic flow. J. Atmos. Sci. **20**, 130–141 (1963)

[LR05] L.E. Lisiecki, M.E. Raymo, A Pliocene-Pleistocene stack of 57 globally distributed benthic δ^{18} records. Paleoceanography **20**, PA 1003 (2005)

[Maa94] L.R.M. Maas, A simple model for the three-dimensional, thermally and wind-driven ocean circulation. Tellus **46A**, 671–680 (1994)

[MOS89] F. Martinelli, E. Olivieri, E. Scoppola, Small random perturbations of finite- and infinite-dimensional dynamical systems: Unpredictability of exit times. J. Stat. Phys. **55**(3/4), 477–504 (1989)

[MPR10] C. Marinelli, C. Prévôt, M. Röckner, Regular dependence on initial data for stochastic evolution equations with multiplicative Poisson noise. J. Funct. Anal. **258**(2), 616–649 (2010)

[MTE99] A.J. Majda, I. Timofeyev, E. Vanden Eijnden, Models for stochastic climate prediction. Proc. Natl. Acad. Sci. USA **96**, 14687–14691 (1999)

[Mue98] C. Mueller, The heat equation with Lévy noise. Stoch. Process. Appl. **74**, 133–151 (1998)

[Myt02] L. Mytnik, Stochastic partial differential equations driven by stable noise. Probab. Theor. Relat. Fields **123**, 157–201 (2002)

[NGR04] Members NGRIP, Ngrip data. Nature **431**, 147–151 (2004)

[Par75] E. Pardoux, Equations aux dérivés partielles stochastiques de type monotone (French). Séminaire sur les Equations aux Dérivées Partielles (1974–1975), III, Exp. No. 2, 10 pp., 1975

[PBL+97] J.R. Petit, I. Basile, A. Leruyet, D. Raynaud, J. Jouzel, M. Stievenard, V.Y. Lipenkov, N.I. Barkov, B.B. Kudryashov, M. Davis, E. Saltzmann, V. Kotlyakov, Four climate cycles in Vostok ice core. Nature **387**, 359–360 (1997)

[PJR+99] J.R. Petit, J. Jouzel, D. Raynaud, N.I. Barkov, J.-M. Barnola, I. Basile, M. Benders, J. Chapellaz, M. Davis, G. Delayque, M. Delmotte, V.M. Kotlyakov, M. Legrand, V.Y. Lipenkov, C. Lorius, L. Pépin, C. Ritz, E. Saltzmann, M. Stievenard, Climate and atmospheric history of the past 420,000 years from the vostok ice core, Antarctica. Nature **399**, 429–436 (1999)

[PR07] C. Prévôt, M. Röckner, *A Concise Course on Stochastic Partial Differential Equations* (Springer, Berlin, 2007)

[Pré10] C. Prévôt, Existence, uniqueness and regularity w.r.t. the initial condition of mild solutions of SPDEs driven by Poisson noise. Infin. Dimens. Anal. Quant. Probab. Relat. Top. **13**(1), 133–163 (2010)

[PXZ11] E. Priola, L. Xu, J. Zabcyk, Exponential mixing for some SPDEs with Lévy noise. Stoch. Dynam. **11**(2–3), 521–534 (2011)

[PZ06] S. Peszat, J. Zabczyk, Stochastic heat and wave equations driven by an impulsive noise, in *Stochastic Partial Differential Equations and Applications VII*, ed. by G. Da Prato, L. Tubaro. Lecture Notes in Pure and Applied Mathematics, vol. 245 (Chapman and Hall/CRC, Boca Raton, 2006), pp. 229–242

[PZ07] S. Peszat, J. Zabczyk, *Stochastic Partial Differential Equations with Lévy Noise (an Evolution Equation Approach)* (Cambridge University Press, Cambridge, 2007)

[PZ10] E. Priola, J. Zabcyk, Structural properties of semilinear SPDEs driven by cylindrical stable processes. Probab. Theor. Relat. Fields **29**(2), 97–137 (2010)

[Rah03] S. Rahmstorf, Timing of abrupt climate change: A precise clock. Geophys. Res. Lett. **30**(10), 1510 (2003)

[Rau02] G. Raugel, Global attractors in partial differential equations, in *Handbook of Dynamical Systems*, vol. 2, ed. by B. Fiedler (Elsevier, Amsterdam, 2002), pp. 885–982

[Rob01] J.C. Robinson, in *Infinite-Dimensional Dynamical Systems: An Introduction to Dissipative Parabolic PDEs and the Theory of Global Attractors.* Cambridge Texts in Applied Mathematics (Cambridge University Press, Cambridge, 2001)

[RZ07] M. Röckner, T. Zhang, Stochastic evolution equations of jump type: Existence, uniqueness and large deviation principles. Potential Anal. **3**, 255–279 (2007)

[Sto61] H.M. Stommel, Thermohaline convection with two stable regimes of flow. Tellus **13**, 224–230 (1961)

[Sto05] S. Stolze, Stochastic equations in Hilbert space with Lévy noise and their applications in finance. Diplomarbeit Universität Bielefeld, 2005

[SY02] S.G. Sell, Y. You, in *Dynamics of Evolutionary Equations*. Applied Mathematical Sciences, vol. 143 (Springer, Berlin, 2002)

[Tem92] R. Temam, in *Dynamical Systems in Physics and Applications*. Springer Texts in Applied Mathematics (Springer, Berlin, 1992)

[Wak06] T. Wakasa, Exact eigenvalues and eigenfunctions associated with linearization of Chafee-Infante equation. Funkcialaj Ekvacioj **49**, 321–336 (2006)

[Wal81] J.B. Walsh, A stochastic model of neural response. Adv. Appl. Probab. **13**, 231–281 (1981)

[Wal86] J.B. Walsh, An introduction to stochastic partial differential equations, in *Ecole d'été de probabilité de Saint-Flour XIV - 1984*. Lecture Notes in Mathematics, vol. 1180 (Springer, Berlin, 1986), pp. 265–437

[Wu10] J.-L. Wu, Lévy white noise, elliptic SPDEs and Euclidian random fields, in *Recent Development in Stochastic Dynamics and Stochastic Analysis*. Interdisciplinary Mathematical Sciences, vol. 8 (2010), pp. 251–268

[Xia92] A.H. Xia, Weak convergence of jump processes, in *Séminaire de Probabiliés XXVI*. Lecture Notes in Mathematics, vol. 1526 (Springer, Berlin, 1992), pp. 32–46

[Xie10] Y. Xie, Poincaré inequality for linear SPDEs driven by Lévy noise. Stoch. Process. Appl. **120**(10), 1950–1965 (2010)

LECTURE NOTES IN MATHEMATICS Springer

Edited by J.-M. Morel, B. Teissier; P.K. Maini

Editorial Policy (for the publication of monographs)

1. Lecture Notes aim to report new developments in all areas of mathematics and their applications - quickly, informally and at a high level. Mathematical texts analysing new developments in modelling and numerical simulation are welcome.
   Monograph manuscripts should be reasonably self-contained and rounded off. Thus they may, and often will, present not only results of the author but also related work by other people. They may be based on specialised lecture courses. Furthermore, the manuscripts should provide sufficient motivation, examples and applications. This clearly distinguishes Lecture Notes from journal articles or technical reports which normally are very concise. Articles intended for a journal but too long to be accepted by most journals, usually do not have this "lecture notes" character. For similar reasons it is unusual for doctoral theses to be accepted for the Lecture Notes series, though habilitation theses may be appropriate.

2. Manuscripts should be submitted either online at www.editorialmanager.com/lnm to Springer's mathematics editorial in Heidelberg, or to one of the series editors. In general, manuscripts will be sent out to 2 external referees for evaluation. If a decision cannot yet be reached on the basis of the first 2 reports, further referees may be contacted: The author will be informed of this. A final decision to publish can be made only on the basis of the complete manuscript, however a refereeing process leading to a preliminary decision can be based on a pre-final or incomplete manuscript. The strict minimum amount of material that will be considered should include a detailed outline describing the planned contents of each chapter, a bibliography and several sample chapters.
   Authors should be aware that incomplete or insufficiently close to final manuscripts almost always result in longer refereeing times and nevertheless unclear referees' recommendations, making further refereeing of a final draft necessary.
   Authors should also be aware that parallel submission of their manuscript to another publisher while under consideration for LNM will in general lead to immediate rejection.

3. Manuscripts should in general be submitted in English. Final manuscripts should contain at least 100 pages of mathematical text and should always include

   – a table of contents;
   – an informative introduction, with adequate motivation and perhaps some historical remarks: it should be accessible to a reader not intimately familiar with the topic treated;
   – a subject index: as a rule this is genuinely helpful for the reader.

   For evaluation purposes, manuscripts may be submitted in print or electronic form (print form is still preferred by most referees), in the latter case preferably as pdf- or zipped psfiles. Lecture Notes volumes are, as a rule, printed digitally from the authors' files. To ensure best results, authors are asked to use the LaTeX2e style files available from Springer's web-server at:

   ftp://ftp.springer.de/pub/tex/latex/svmonot1/ (for monographs) and
   ftp://ftp.springer.de/pub/tex/latex/svmultt1/ (for summer schools/tutorials).

Additional technical instructions, if necessary, are available on request from lnm@springer.com.

4. Careful preparation of the manuscripts will help keep production time short besides ensuring satisfactory appearance of the finished book in print and online. After acceptance of the manuscript authors will be asked to prepare the final LaTeX source files and also the corresponding dvi-, pdf- or zipped ps-file. The LaTeX source files are essential for producing the full-text online version of the book (see http://www.springerlink.com/openurl.asp?genre=journal&issn=0075-8434 for the existing online volumes of LNM). The actual production of a Lecture Notes volume takes approximately 12 weeks.

5. Authors receive a total of 50 free copies of their volume, but no royalties. They are entitled to a discount of 33.3 % on the price of Springer books purchased for their personal use, if ordering directly from Springer.

6. Commitment to publish is made by letter of intent rather than by signing a formal contract. Springer-Verlag secures the copyright for each volume. Authors are free to reuse material contained in their LNM volumes in later publications: a brief written (or e-mail) request for formal permission is sufficient.

Addresses:

Professor J.-M. Morel, CMLA,
École Normale Supérieure de Cachan,
61 Avenue du Président Wilson, 94235 Cachan Cedex, France
E-mail: morel@cmla.ens-cachan.fr

Professor B. Teissier, Institut Mathématique de Jussieu,
UMR 7586 du CNRS, Équipe "Géométrie et Dynamique",
175 rue du Chevaleret
75013 Paris, France
E-mail: teissier@math.jussieu.fr

For the "Mathematical Biosciences Subseries" of LNM:

Professor P. K. Maini, Center for Mathematical Biology,
Mathematical Institute, 24-29 St Giles,
Oxford OX1 3LP, UK
E-mail : maini@maths.ox.ac.uk

Springer, Mathematics Editorial, Tiergartenstr. 17,
69121 Heidelberg, Germany,
Tel.: +49 (6221) 4876-8259

Fax: +49 (6221) 4876-8259
E-mail: lnm@springer.com